Tucholsky Wagner Zola Scott Sydow Freud Schlegel
Turgenev Wallace Fonatne
Twain Walther von der Vogelweide Fouqué Friedrich II. von Preußen
Weber Freiligrath
Fechner Kant Ernst Frey
Fichte Weiße Rose von Fallersleben Richthofen Frommel
Engels Fielding Hölderlin
Fehrs Faber Eichendorff Tacitus Dumas
Flaubert Eliasberg Ebner Eschenbach
Feuerbach Maximilian I. von Habsburg Fock Eliot Zweig Vergil
Goethe Ewald London
Mendelssohn Balzac Shakespeare Elisabeth von Österreich Ganghofer
Lichtenberg Rathenau Dostojewski Gjellerup
Trackl Stevenson Doyle
Mommsen Tolstoi Hambruch
Thoma Lenz Hanrieder Droste-Hülshoff
Dach von Arnim Hägele Hauff Humboldt
Reuter Verne Rousseau Hagen Hauptmann
Karrillon Garschin Gautier
Damaschke Defoe Hebbel Baudelaire
Descartes Hegel Kussmaul Herder
Wolfram von Eschenbach Dickens Schopenhauer Rilke George
Bronner Darwin Melville Grimm Jerome Bebel
Campe Horváth Aristoteles Proust
Bismarck Vigny Barlach Voltaire Federer Herodot
Gengenbach Heine
Storm Casanova Tersteegen Grillparzer Georgy
Chamberlain Lessing Langbein Gilm
Brentano Lafontaine Gryphius
Strachwitz Claudius Schiller Kralik Iffland Sokrates
Katharina II. von Rußland Bellamy Schilling
Gerstäcker Raabe Gibbon Tschechow
Löns Hesse Hoffmann Gogol Wilde Gleim Vulpius
Luther Heym Hofmannsthal Klee Hölty Morgenstern Goedicke
Roth Heyse Klopstock Puschkin Homer Kleist
Luxemburg La Roche Horaz Mörike Musil
Machiavelli Kierkegaard Kraft Kraus
Navarra Aurel Musset Lamprecht Kind Kirchhoff Hugo Moltke
Nestroy Marie de France
Nietzsche Nansen Laotse Ipsen Liebknecht
von Ossietzky Marx Lassalle Gorki Klett Ringelnatz
May vom Stein Lawrence Leibniz
Petalozzi Puschkin Irving
Platon Knigge
Sachs Pückler Michelangelo Kock Kafka
Poe Liebermann Korolenko
de Sade Praetorius Mistral Zetkin

The publishing house tredition has created the series **TREDITION CLASSICS**. It contains classical literature works from over two thousand years. Most of these titles have been out of print and off the bookstore shelves for decades.

The book series is intended to preserve the cultural legacy and to promote the timeless works of classical literature. As a reader of a **TREDITION CLASSICS** book, the reader supports the mission to save many of the amazing works of world literature from oblivion.

The symbol of **TREDITION CLASSICS** is Johannes Gutenberg (1400 – 1468), the inventor of movable type printing.

With the series, tredition intends to make thousands of international literature classics available in printed format again – worldwide.

All books are available at book retailers worldwide in paperback and in hardcover. For more information please visit: www.tredition.com

tredition was established in 2006 by Sandra Latusseck and Soenke Schulz. Based in Hamburg, Germany, tredition offers publishing solutions to authors and publishing houses, combined with worldwide distribution of printed and digital book content. tredition is uniquely positioned to enable authors and publishing houses to create books on their own terms and without conventional manufacturing risks.

For more information please visit: www.tredition.com

The Wonders of the Jungle Book One

Sarath Kumar Ghosh

Imprint

This book is part of the TREDITION CLASSICS series.

Author: Sarath Kumar Ghosh
Cover design: toepferschumann, Berlin (Germany)

Publisher: tredition GmbH, Hamburg (Germany)
ISBN: 978-3-8491-5088-4

www.tredition.com
www.tredition.de

Copyright:
The content of this book is sourced from the public domain.

The intention of the TREDITION CLASSICS series is to make world literature in the public domain available in printed format. Literary enthusiasts and organizations worldwide have scanned and digitally edited the original texts. tredition has subsequently formatted and redesigned the content into a modern reading layout. Therefore, we cannot guarantee the exact reproduction of the original format of a particular historic edition. Please also note that no modifications have been made to the spelling, therefore it may differ from the orthography used today.

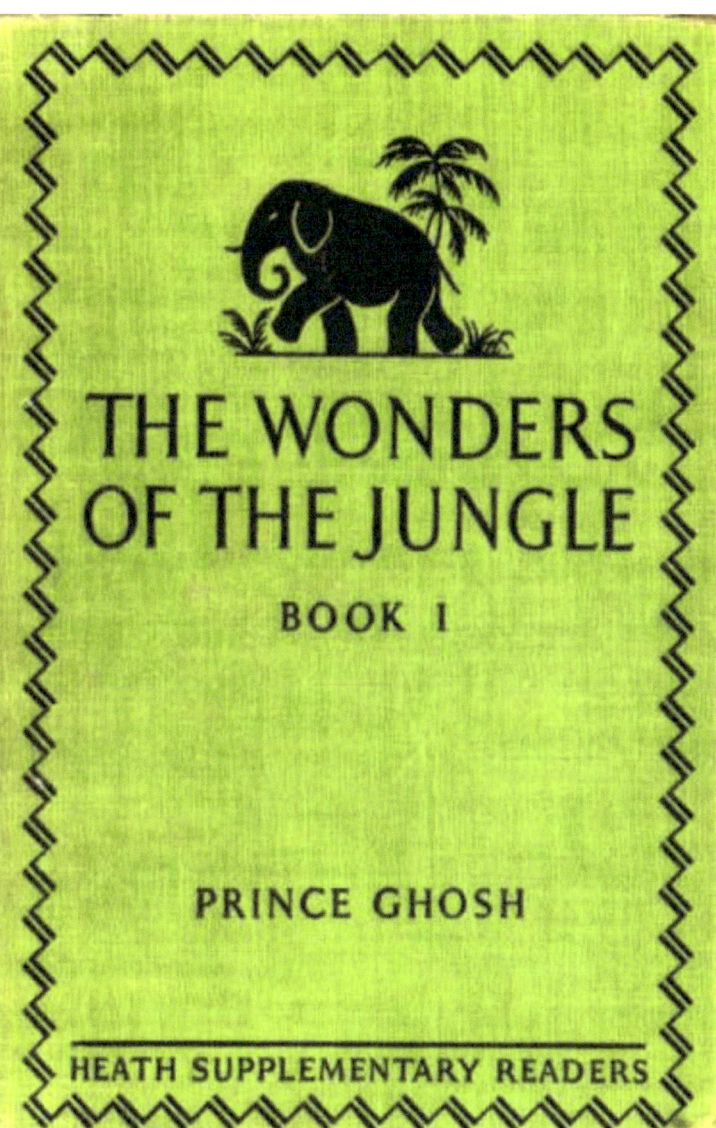

THE WONDERS OF THE JUNGLE

BOOK I

PRINCE GHOSH

HEATH SUPPLEMENTARY READERS

Midnight Pool: Animals Drinking

HEATH SUPPLEMENTARY READERS

THE WONDERS OF THE JUNGLE

PRINCE SARATH GHOSH

BOOK ONE

D. C. HEATH AND COMPANY
BOSTON   NEW YORK   CHICAGO   ATLANTA
DALLAS   SAN FRANCISCO   LONDON

PRINCE SARATH GHOSH

Book One

PREFACE

One of the great thinkers of the world has said that all the sciences are embodied in natural history. Hence natural history should be taught to a child from an early age.

Perhaps the best method of teaching it is to set forth the characteristics of animals in the form of a narrative. Then the child reads the narrative with pleasure and almost as a story, not as a tedious "lesson."

I have followed that method in the Wonders of the Jungle. The present work (Book One) is intended to be a supplementary reader for the earlier grades in grammar schools. If it be found useful, I shall write one or two more books in progressive order for the use of higher grades.

In Book One I have depicted only such wild animals as appeal to the interest of young children, and even to their sympathy and love. In subsequent books I shall describe the animals that prey upon others. As those animals are not lovable, it would be better for the child to read about them a year or two later. But even to those animals I shall be just, and shall depict their good qualities as well as their preying habits. How many people know that the very worst animal, the tiger, is a better husband and father than many men? Or that the ferocity of the tigress is prompted entirely by her maternal instinct—and that in every case of unusual ferocity yet recorded it was afterward found that there was a helpless cub somewhere near? Hence in subsequent books I shall enter more fully into the causes of animal instincts and characteristics—their loves and their hates and their fears. [iv]

Regarding the scheme of Book One, the animals are described in their daily life, and the main scientific facts and principles concerning each animal are woven into the narrative as a part of that daily life. But while teaching science to the child in that pleasant form, a few other purposes have also been kept in view:—

1. To cultivate the child's imagination. True imagination is the ability to visualize mentally the realities of life, not what is unreal—for which it is so often mistaken. Hence in this book the child is

helped to visualize the animals in their actual haunts, and to see each incident as it actually happens.

2. To cultivate the child's reasoning faculty. The child is encouraged at every step to think and to reason why the animal does certain things; *e.g.* why the elephant does not drink directly with its mouth, but has to squirt the water into it with the trunk.

3. To teach a moral from the study of animals. The whole of Creation is one immense and beautiful pattern: so the child may well be trained to see the pattern in this also. And as a practical benefit from the study of animals, the child may learn thereby the value of certain qualities, such as obedience, discipline, and good citizenship— *e.g.* as in the remarkable case of the elephant, the buffalo, and the flamingo, as described in the text. In this regard I have kept in mind the very useful suggestions formulated a few years ago by the Moral Education League of Great Britain, under the patronage of Queen Mary, five of whose children at that time ranged in age from seven to fifteen. One of the functions of education is to present to the child the noblest and the most elevated of ideals. I have sought to do that in almost every chapter.

I have to acknowledge my obligation to the New York Kindergarten Association for its valuable cooperation in putting this book through a practical test. The Kindergarten Association on more than one occasion provided me with a large audience of children, ranging in age from six to nine, ex-pupils of the Association, who are now in the public schools.

CONTENTS

[v]

	CHAPTER	
I.	The Midnight Pool	
	Elephants Drink First—But Down Stream	
	How the Elephant Drinks	
	Why the Elephant Drinks with his Trunk	
II.	The Law of the Jungle	
	How Buffaloes Come to Drink—In Rows	
	Buffalo Knights Guard the Timid Deer	
	Wild Pigs—Careless	
	Red Dogs—Bold, Fearing Nobody	
	Other Animals Come Alone	
	The Law of the Jungle—Clear Water for All	
III.	The Elephants' Bath	
	Elephant Child Obeys Mamma—or Gets Spanked	
	How the Elephant Child is Bathed	
	How the Elephant Child Learns to Swim	
IV.	Elephants: The Tricks of the Jungle	
	Elephant Child Learns to Feed	
	Elephant Child "Swats" Tormenting Flies	
	Elephant Covers Back from Hot Sun	
	How Elephants Walk under Water	
	How Elephants Break Down or Pull Out Trees	
V.	Elephants: The Tricky Trap	
	The Elephant Taps Suspicious Ground with his Trunk	
	Elephant Tricks the Tricky Trappers	

VI. Buffaloes: The Knights of the Jungle
> Buffaloes Cover Body with Mud against Flies
>
> How Buffaloes Guard against Tiger while Feeding
>
> How Buffaloes Know Danger is Coming — Three ways
>
> Buffalo Sentinels
>
> Buffaloes Make a Ring when Tiger Comes
>
> Small Animals Find Safety in Buffalo Ring

VII. Taming the Buffalo
> Wild Buffaloes Tamed Quickly by Kindness
>
> Little Boys Take Charge of Buffaloes
>
> How the Big Buffaloes Love the Little Boys

VIII. The Buffalo and the Boy

IX. Deer and Antelope
> Horns and Antlers Different in Three Ways
>
> Elk and Other American Deer
>
> Other Kinds of Deer
>
> Barking Deer — One of the Wonders of Nature

X. Deer and Antelope: Their Special Gifts
> Each Animal has the Gift he Needs Most

XI. The Camel
> The Camel's Wonderful Gifts

XII. The Camel and the Thief

XIII. Bears
> The Polar Bear
>
> American Bears
>
> Other Bears

XIV. Bears: The Tricky Trap

XV. Bright Birds
- The Flamingo
- The Parrot
- The Cockatoo
- The Peacock
- The Golden Pheasant
- The Snowy Egret

XVI. The Caged Parrot

ILLUSTRATIONS

Midnight Pool: Animals Drinking

The Buffaloes and the Blue Deer

An Elephant Giving himself a Shower Bath.

An Elephant Mamma Carrying her Child across the River

An Elephant Breaking a Tree with his Foot

Elephant Pulling Bananas out of a Tricky Trap

The Buffalo that Lives in India

The Tiger and the Ring of Buffaloes

Tame Water Buffaloes Plowing in the Rice Fields

Antelope

Elk

Arabian Camel — with One Hump

Bactrian Camel — with Two Humps

Sand Storm in the Desert

Crossing the Desert with Camels

Polar Bear

Himalayan Black Bear

A Bear Fighting a Block of Stone

A Flamingo Colony

Snowy Egrets

[1]

THE WONDERS OF THE JUNGLE

CHAPTER I

The Midnight Pool

My dear, I shall tell you all about the wonders of the jungle. You have seen many animals in the zoo or in a circus—elephants, bears, lions, tigers, leopards, and many others. But the jungle is the place where these animals live before they are brought to the zoo or the circus.

In fact, *jungle* really means a *wild place*; that is, a place where trees and bushes grow quite wild, so that men never cut down the trees or clear away the bushes. That is the natural home for all sorts of animals.

Now I am going to tell you about the wonderful way in which they live there with their families, as we do in our homes; for the Papas and Mammas among the animals are just as fond of their children as ours are. So you must *imagine* [2] that you are going into the jungle with me, so that I can show you everything. You see, it is just like a game of *pretending*, that we are going to play.

There is actually a place in the jungle where you can see all the animals at once. In fact, that place is so wonderful that King George and Queen Mary of England went to see it; that was a few years ago, when they went to India, which is a far-away country. For in India there is a huge jungle where many thousands of animals live.

So you must *pretend* that I am taking you to the Royal party, and that you are sitting with the King and Queen and all the fine men and lovely ladies; and we are watching the animals, while I tell you all about them.

First, I must tell you that it is midnight, and all the animals are coming to a stream of water to drink. This stream is a river about twice as wide as a large street in your home town. We are sitting on the bank, on one side of the stream; and the animals are coming to drink on the bank on the other side.

"But," you may say, "will not the animals see us across the stream, and get frightened and run away?" [3]

That is quite true. But the King and Queen had thought of that. So they ordered a lot of men to put a large net on their side of the stream, just in front of them, and then to cover the net with twigs and leaves so cleverly that the animals thought the leaves were a part of the jungle, and did not see the people on the other side of the net.

So the King and Queen, and you and I, can peep quietly through the leaves and watch the animals. Almost all wild animals drink at midnight; so we shall see them now.

Where will the animals come from? You see the stream before us; well, on the other side of it is the jungle, where the animals live. Right in front of us we see a gap in the jungle close to the bank. That gap was made by *elephants* by beating down the bushes with their feet. They made it long ago to come to the water, and now they use it every night. In fact, it is known among the jungle folks as the *Elephant Path*; for no other animal would dare to use it before the elephants did.

The elephants, being the biggest of all animals, are the *lords of the jungle*; so they have the right to come first to drink. They are also the wisest of all animals. You have seen many kinds [4] of animals—elephants, horses, dogs, monkeys, and others—do funny tricks in a circus. Now, all these animals except the elephant have to be *taught* to do tricks; the elephant is the only animal that can think out a trick for itself.

Of course in a circus there is always a teacher, or trainer, to show even the elephant how to do tricks; but in the jungle the elephant can find out how to do things for itself.

Very soon I shall tell you about the tricks which the elephant actually does in the jungle; and as you hear about them, you must *think*! Why? Because then you will know *why* the elephant does these things—and that will show you how clever *you* can be!

Elephants Drink First—but Down Stream

First let us watch the elephants as they come to the river through the gap in the jungle.

See! They come one at a time, *one behind another;* for the gap is not big enough for more than one at the same time. The elephant is so big that it can get through the jungle only in this way.

First come a number of *bull elephants*. They are the Papa elephants; you can always tell them by the *huge tusks* they have. The bulls [5] come first, in case there are any enemies waiting to hurt their children; for then the bulls can drive off the enemies.

As each bull elephant comes through the gap, you see him turn to our right, which is *down* the stream—that is, the way the water flows. You see the first one walk along the bank that way, and the second comes after him, then the third, and so on.

But why do they walk along the bank? To make room, of course, for all their friends who are still coming from behind. In this way about a dozen bull elephants come ahead of all the others.

After them you see the *cow elephants*, also in a line, one behind another. They are the Mamma elephants; and nearly every one of them has a baby elephant trotting in *front* of her. You have often seen the ordinary cow that gives you milk; when she goes to graze in the field, her baby, or calf, trots by her side.

But the Mamma elephant is much wiser, and always tells her baby to toddle in *front* of her, in case any one comes suddenly to hurt or steal the baby. For a tiger sometimes wants to pounce on the baby from the side, grab it quickly, and carry it away. But he cannot do it if the [6] baby is right in front of its Mamma; for then she will drive him off with her tusks, even if they are not quite so big as the tusks that the Papa elephants have.

As the Mammas reach the bank, each with her baby, you see them also walk along the bank down stream in a long line.

After all the Mammas and babies have come, you see another set of bull elephants coming out of the jungle. Why? Because some enemy might try to attack the Mammas and the babies from the *back*; so these bull elephants are there to guard them. You see, the Mammas and the babies are *always in the middle*, safe from all harm.

When all the elephants have reached the stream, they stand in line and face the water. All these elephants belong to *one herd*; you can count about a hundred. A herd of elephants is really a *republic*, like the United States of America, and has a President, who is the wisest bull in the herd.

In another book I shall tell you how the elephants choose their President, and make laws, and keep order in the herd; how they choose some strong bulls among them to act as *policemen* in the herd, and catch and punish [7] any naughty elephant who becomes a *rogue*; and how, if two elephants start quarrelling and fighting like naughty boys, the police elephants have to catch and punish both of them. Also, I shall tell you how the President has to lead the herd every day when they go in search of food, so that they will have plenty to eat.

And in the jungle, as there are other elephant herds and sometimes two herds find the same feeding ground, and then start quarrelling and fighting as to who found it first, it is the duty of the President to keep his own herd away from the two that are fighting, and not mix in the fight in any way. All these wonderful things and many others you will read in the other book, when you are a little older.

But let us see what the President has to do when the whole herd is standing in line, facing the water. He is at the bottom of the line, far down stream; so he looks up along the line to see that all are ready. Then he gives the signal for them to begin drinking; he does this by dipping his trunk into the water. Then the second one sees him do it, and does the same; in that way each elephant higher up the line sees that the next one below him has started [8] drinking, so he too does the same. Soon they are all drinking, as you see in the picture at the beginning of this book.

But why does the President have to give the signal to begin? Why is it that any elephant, anywhere along the line, cannot start drinking, just as he or she pleases? Think!

Because if any one along the line started drinking too soon, he might muddy the water for those that stood *below* him along the line, because the water flows down that way. But if the lower ones drank a little before, it would not matter if they *did* muddy the water, for the higher ones would still have clear water to drink. That is why the lowest one drinks first, then the next, and so on up the line. Is not that very wise, and very fair to all?

How the Elephant Drinks

But you must not think that an elephant actually drinks *through* his trunk! He does not! The elephant's trunk is really his nose, though it is a very long nose. What he does is to dip the trunk into the stream and suck in the water about halfway up the trunk; then he curls up the tip of the trunk and gets it near his mouth; then he *blows* through the nose, and [9] *squirts* the water into his mouth. Of course he has to do that many times, to get enough to drink. But he tries each time to dip only the tip of the trunk into the stream, so as not to muddy the water willfully!

Why the Elephant Drinks with His Trunk

But, you may say, why cannot he drink like other animals, by going right into the stream till he gets his *mouth* into the water? Because his mouth is so high up, and his neck is so stiff, that he would have to go quite two or three yards deep into the stream before he could get his mouth into the water, and then his heavy feet would stir up the mud in the stream where he was standing, and so dirty the very water he was drinking.

Now you see what a wise animal the elephant is! The only way he could get clear water to drink was by having a long nose! And that is exactly what happened many, many years ago—his nose became long enough to reach the water from the bank. How that happened I shall tell you in another book, as you will not understand it till you are a year or two older.

All the grown-up elephants drink in this way, [10] and also some of the elephant children whose trunks have grown long enough to reach the stream. But what about a baby elephant? Why, its Mamma fills her own trunk with water, puts the tip into the baby's mouth and squirts the water into it.

But now after watching the elephants—who are on our right, down the stream—let us turn our eyes to the left, and look *up* the stream.

[11]

CHAPTER II

The Law of the Jungle

Hush! Here come all the animals! The *buffaloes*, the *blue deer*, the *red deer*, the *wild pigs*, the *hyenas*, the *wolves*, the *red dogs*, and many others. Watch and see how each kind of animal comes; it is not always in the same way. The moon is now shining clear above the trees, and we can see a long way up the stream.

See the *buffaloes!* They come a little *above the elephants*. But they do not come one behind another in a line, like the elephants. They come three or four together. They also have beaten down the bushes there years ago, to make a drinking place; and it is wide enough for three or four of them to drink at the same time, side by side.

How Buffaloes Come to Drink — in Rows

But why must they drink three or four at the same time? Because the buffaloes are like a body of soldiers, one row behind another. Sometimes twenty or thirty rows make up a [12] herd. We see only the first row drinking now, but soon we shall see the others behind.

And why do the buffaloes come like a body of soldiers? Because they are afraid of their enemy — the tiger! Once upon a time the buffaloes lived scattered about, and many of them got eaten by the tiger, one at a time. Then those that escaped from the tiger became wise; they joined together like a body of soldiers, so that they could beat off the tiger. How they came to do that, I shall tell you at another time.

But now let us watch the first row drinking. They are all *bull buffaloes*, the Papas of the herd; you can tell that by their *huge horns*, a yard long on each side of the head. You see how the buffaloes stand side by side, so that their horns almost touch one another. That is the way the buffaloes have marched to the stream from their feeding place — horn to horn. Why? Because no prowling tiger can get past those horns.

Watch the first row as it finishes drinking; the whole row wheels around to the side like soldiers. Then the buffaloes that have had their drink march to the back of the herd, and stand there in a row facing the jungle. [13]

Meanwhile the second row in the front has stepped to the water to drink. These also are bull buffaloes. When they finish drinking, they also wheel, march to the back of the herd, and there stand behind the first row. In this way four or five rows of bulls drink, one after the other, and go to the back of the herd.

Next come about a dozen rows of *cow buffaloes* and their calves, or children. You see again, like the elephants, the Mammas and children among the buffaloes are also *in the middle*, safe from all harm.

Then at the end there are four or five rows of bull buffaloes again, to guard the Mammas and the children from enemies in the back.

Buffalo Knights Guard the Timid Deer

But wait a moment! Before the buffaloes go away, a most wonderful thing happens. You have read stories, how once upon a time there were brave knights who used to come to the help of ladies who were in danger. Well, you will be glad to know that these bull buffaloes are just like those brave knights. Do you see that timid little shadow creeping in by the side of the buffaloes?

She is a *blue deer*, a very timid lady indeed; [14] for she knows that a tiger is waiting in the high ground behind, to catch her. It is the last chance of the tiger to get his supper; so he waits by the high ground behind, and watches for some weak animal like the deer to come to drink.

But the blue deer knows that; so she hides in the bushes, and waits for the buffaloes to come to drink. Then as the buffaloes come to the water, row after row, horn to horn, she tries to creep in toward them; she even tries to creep in *under* the horns of the buffaloes, knowing that there she will be quite safe from the tiger. It takes her a long time to reach the buffaloes in that way, without being caught by the tiger.

But do you see the wonderful thing? The buffaloes wait a little for her! They take a little longer to drink, to give her a chance to reach the water by their side. Like the brave knights, they feel proud of helping a lady.

Now see! The blue deer also has finished drinking. She goes away with the buffaloes, under their horns. They all reach the jungle again. She looks carefully: the tiger is watching her, but he dares not come too near. She sees where he is—then suddenly she gives a [15] leap—another leap—and another—quickly! The tiger leaps after her—but she leaped first! She is gone! She is safe!

The Buffaloes and the Blue Deer

[17]

The tiger is furious. He stands a moment before the buffaloes, growling with rage. But the bulls in front of the herd paw the ground, and rattle their horns with one another. They are going to charge!

But that tiger does not wait for the charge of the bull buffaloes. He does not want to be trampled into a mess under their hoofs, or cut up into pieces with their horns. Instead, he sneaks away, growling. He sneaks back to the stream, to wait for some other weak animal.

So, you see, the jungle folks are in many ways just like us; for a brave man always helps a lady or anybody who needs his help.

But now let us watch the stream higher up.

Wild Pigs—Careless

Here come the *wild pigs*. They are not exactly a herd; but still there are many dozens of them, all one large family with all their relations—cousins and uncles and aunts. Some of the wild pigs are called *boars*; they are the Papas among the wild pigs. You can always tell them by the two *sharp tusks*, or teeth, one [18] on each side, which grow *upward* from their under jaw. Each tusk is as long as a knife, and so sharp that a tiger does not always care to fight with a boar.

The wild pigs drink in any fashion, and go off in any fashion—just as they like. They trust to luck or to the sharp tusks of some of the boars to guard them from danger. But they have not learned enough yet to do things in proper order.

Red Dogs—Bold, Fearing Nobody

Meanwhile other animals have also come. The moon is now quite high in the sky. A band of shadows in the moonlight seems to fall upon the water. It is a pack of *red dogs*; they have come boldly, as they are afraid of nothing. For if a hungry tiger attacks them, the whole pack will jump on the tiger and tear him down—that is, the tiger could kill dozens of the dogs in a few minutes, but then the rest of the wild red dogs would tear the tiger to pieces.

So the red dogs are not afraid as they come flocking to the stream. They lap up the water with their lolling tongues. Then they look up at the moon. Do you see what they are doing? Can you *hear* them? They are *howling at the* [19] *moon in a chorus*. Dogs always howl at the moon. Men do not quite know just why dogs do that. But perhaps they do it because they are glad and satisfied, and are trying to *sing!* When *you* sing, and there is a dog near by, you may hear him start howling. He does that, I suppose, because he likes your singing, and wants to join in the chorus!

So the wild dogs of the jungle also howl when they are glad. Then, after the red dogs have howled as long as a song, they scamper off into the jungle again. That shows, I suppose, that their howling was really a song!

Other Animals Come Alone

The red dogs are the last of the animals that come in a bunch. Now you see other animals coming one by one. A sneaking shadow there! It must be a *hyena*. That is an animal that eats what remains from some other animal's supper; so the hyena waits to see if a tiger or a leopard has caught any supper, or else it will have to go hungry.

But hush! Here is a *red deer* coming carefully to the water. This animal is much bigger than the blue deer, and more able to take care of herself. But, still, she comes very quietly, [20] looking to right and left to make sure that the tiger is not just in that place. She reaches the water and starts drinking. But do you see how her ear is bent to the side? The red deer is listening most carefully, even while she is drinking!

But look, look! The bush behind the deer parts very slowly, and a huge yellow form crouches there! It is the tiger!

He is not near enough to jump on the deer; so he takes one step forward — as softly as a cat!

But the deer has heard the footfall! For she can hear even a leaf when it falls to the ground. And in that one second, even while she was drinking, the red deer has turned and leaped to the side. The tiger has also leaped at the same time, and he aimed at the place where the deer *was*. But the deer has just left that place, and the next second she gives another leap, like a flash, and gets out of the tiger's reach.

The tiger stands where he leaped, and growls with rage. He knows it would be no use chasing the deer, as *the deer can run much faster*. So he stands there, and growls for quite a while. Then, as he did not get any supper that night, [21] he can at least have a drink. So he drinks and goes away, still growling.

Now all is quiet at last at the midnight pool, as all the animals have gone away.

The Law of the Jungle — Clear Water for All

But before *we* leave the place, I want you to remember something. I showed you first the elephants; they were on our right — that is, *down* the stream, the way the water flows. And the elephants drank first among all the animals.

Then all the other animals came to the stream, but more to our left — that is, *up* the stream. Why was that? Think!

I shall tell you. By the time the elephants finish drinking by dipping their trunks into the stream many times, the water begins to get muddy. In fact, after drinking, the elephants jump into the water to have a bath and a swim, as I shall tell you in the next chapter.

So the water gets muddy near the elephants and all the way down stream from that place, as the water flows that way. And as the other animals do not want muddy water to drink, they always go *up* the stream, where the water is still clear. [22]

That is *The Law of the Jungle*, though it is not written down in a book, like the laws among men. The Law of the Jungle says that as the elephants are the lords of the jungle, they shall drink *first*: but they must be careful to drink *down the stream*, so that all the other animals may have a place higher up, where they can get *clear water to drink*.

And that law has never been broken, for many thousands of years, among all the different sorts of animals.

But with men the laws among the different sorts of people, called nations, are often broken, because some of them want all the best things and the best places, and do not care if they muddy the water that their neighbors have to drink.

So, my dear children, we can learn many things from the animals, even how to be better men and women when we grow up.

[23]

CHAPTER III

The Elephants' Bath

I have just told you that, after drinking, the elephants jump into the water and have a bath and a swim. That is, all the grown-up elephants do that, while the little ones stay on the bank and play about.

But, you may ask, why does not the tiger try to grab one of the little ones then? Because even when the Mammas go into the water they keep their eyes on the babies, who play quite near by, so that the Mammas can come to them any minute.

And the Mammas can *smell* a tiger a little before he gets there, so that they have enough time to climb out of the water. Besides, the babies themselves can smell the tiger when he is coming; then they call out to their Mammas by making a queer rumbling sound in their throats, and the Mammas come to them at once, before the tiger can get there.

So all the grown-up elephants can go into the [24] water, without any worry. And at first they have a regular shower bath.

How do they do that? Why, each elephant fills his trunk with water; then he curls up the trunk in the air over his head and squirts the water out, and it falls in a shower all over his body. You can see how he does it in the picture. All the grown-up elephants do that, and even those that are half grown.

After the shower bath, they swim about; but the Mamma elephants do not do that. Why? Because they have to get busy and bathe their little children. They call to the children to stop playing, and come and have a bath—just as our Mammas do.

How do they call? Why, I must tell you at once that all kinds of animals have *a language of their own*. They do not speak exactly as we do, but make different sounds through their mouth or nose, and each sound *means* something.

If the Mamma elephant wants to say "Come here," she makes one kind of sound, and the baby elephant has learned to know exactly what that means. And if the Mamma elephant wants to say "Keep still," she makes another kind of sound, and the baby knows also what that means. [25]

An Elephant Giving himself a Shower Bath

[27]

In this way all animals can talk among themselves. Of course they cannot say many things, as we do, but quite enough to tell what they want.

So each Mamma elephant calls to her child to come and stand on the bank. Now, many of our children often hate to be bathed; and the elephant children are just the same! In fact, the very small ones actually cry and shriek, just like our babies!

Elephant Child Obeys Mamma—or Gets Spanked

But when the Mamma elephant calls to the baby to stop playing and come and stand by the bank, the baby comes at once, even though it hates to be bathed. The baby elephant obeys its Mamma almost the first time, whatever she tells it to do.

But if the baby does not obey, does its Mamma spank it? Of course she does—like all Mammas! The elephant Mamma does the spanking with her trunk.

But I must tell you at once that an elephant child never gets spanked more than once in its life—and that is enough! And some are so good that they *never* get spanked! [28]

The elephant child learns very quickly to obey its Mamma and Papa, and afterwards its trainer or teacher. The elephant child even obeys the very minute it is told to do anything; in fact, sometimes in the jungle there is a sudden danger, even if the elephant child does not see the danger. But its Mamma or Papa sees it.

Then the Mamma or Papa calls out to the child to stop, or come away, or do something, *at once*; and if the child does not do it at once, it may get killed. Among men folks, if a child runs out into the street, and an auto or a street car comes suddenly, then if the child will not obey its Mamma at once and do exactly as she says, the child may be run over and killed. In the jungle the elephant child also has sudden dangers like that, though in a different way.

In the next chapter I shall tell you a wonderful story about a boy elephant who escaped a great danger because he obeyed his Papa at once.

But sometimes it happens that a boy elephant is really naughty — just like a bad boy among men. As you know, a bad boy among men usually grows up to be a bad man, and then he gets into a lot of trouble. In the [29] elephant herd it is just the same; a bad little elephant grows up to be a bad big elephant; it is then called a *rogue*. In another book I shall tell you how the President of the herd orders all the police elephants to stand in a ring around the rogue and give him a most awful spanking. And they do that, not with their trunks this time, but with their *tusks* — which hurt most dreadfully.

How the Elephant Child is Bathed

But now I shall tell you about the baby elephant when its Mamma calls it to come and be bathed. It comes to the edge of the bank, and stands facing its Mamma. Then the Mamma fills her trunk with water, brings the trunk quite near the baby, and squirts the water all over it.

The baby may howl and jump about and make faces, but it *never runs away*! Again and again the Mamma squirts the water, till all the mud and dust of the jungle is washed away from the baby's body. Then she tells the baby to play about on the bank again, while she attends to the bigger children.

What has she got to do to them? She must teach them to swim! [30]

Of course *all animals with four legs know how to swim naturally*; their bodies float in the water quite easily, and they have only to work their legs to move along in the water. But with elephants it is a little different. Why? Just think!

I shall tell you. Although they can float quite naturally, their noses point downward right into the water. As I said before, the elephant's trunk is its nose — that is, the elephant has to *breathe through the trunk*. So of course, if in trying to swim a little elephant kept its trunk down in the water, it would not be able to breathe at all, and would die.

That is why the Mamma elephant has to teach her child how to swim properly. And the way she does it is quite wonderful.

I must first tell you that the trunk is not only like a nose to the elephant, but also is useful as a *hand*; the elephant can hold a lot of things with it, and can even pick up with its tip a tiny thing as small as a pin.

How the Elephant Child Learns to Swim

So the Mamma elephant stretches out her trunk before her, just like an arm, and tells her child to lie across it. In that way she holds up [31] the child in the water, so that the little elephant has only to think of curling up the tip of its own little trunk out of the water to breathe. Then she tells her child to kick out with its legs, so as to move forward through the water.

But sometimes, in kicking out, the little elephant forgets to hold up the tip of its trunk out of the water at the same time; then down goes its trunk into the water, and it cannot breathe!

Then what happens? The Mamma elephant can do nothing, as she is already using her own trunk to hold up her child. So, what is to be done?

Really, the elephants are so wise that they take no chances of that happening. The Papa elephant takes care of that. When he sees that the Mamma is teaching the little elephant how to swim, he always comes near them. He may be swimming about, as if he were enjoying himself; but he is really watching them all the time.

And if the little elephant forgets to hold up its trunk out of the water, the Papa comes quickly, and with one upward stroke of his own trunk he lifts up the little elephant's trunk clear out of the water. Is not that very wise and thoughtful of the Papa elephant? [32]

In that way the little elephant soon learns to do *both* things—that is, to kick out with its legs so as to move along, and also to hold up its trunk to breathe. And then, of course, it can swim properly.

And yet the elephants are so very wise that they never take the risk of tiring out a little elephant, if they have to swim a very long way. Sometimes a whole herd of elephants has to swim across a very wide river. Then the Mamma elephant tells her child to swim in front of her, while she encourages the child from behind with many fond words.

But sometimes after swimming halfway across the river—

"Mamma, I am getting tired!" cries the little one.

"Then come on my back, darling!" says the Mamma.

She dives, and comes up right under the little elephant; so now her child sits on her back. In that way she swims along, and carries her child across the wide river, as you see in the picture. [33]

An Elephant Mamma Carrying her Child across the River

[35]

CHAPTER IV.

Elephants: The Tricks of the Jungle

Now I shall tell you how a little elephant learns all the tricks of the jungle from its Mamma and Papa. By the tricks of the jungle I mean all the things that an animal has to learn in order to get enough to eat every day, what to do when food is scarce, how to be comfortable and happy, and also how to escape from every danger; in fact, these things are very much like what men have to learn, only in a different way.

But the animal folks are better off in one way: what they have to learn is not like a lesson in school, but just play. In fact they learn everything by just playing it as a game! I shall tell you how.

When a baby elephant is quite small, its Mamma has to feed it with milk. Afterwards, when it has teeth, she teaches it to feed from the jungle. All elephants eat tender shoots, herbs, and fresh young leaves; they seize a bough with the trunk, and pull it down in such [36] a way that the end of the bough reaches right into the mouth.

Elephant Child Learns to Feed

First, the Mamma elephant eats like that from several boughs, while the little elephant watches her do it. Then she looks at a low bough within easy reach, and says in the elephant language, "Eat that!"

The little one looks at the bough, grabs it anyhow with its trunk, and pulls it down. But it cannot get the end of the bough *into its mouth*! Instead, the bough pokes it on the forehead, or eyes, or cheeks.

"Hold it straight!" says Mamma, laughing.

The little one tries several times, but still it cannot get the bough to come right. Then its Mamma puts her own trunk over that of her child, and turns it to right or left, till the bough comes exactly into the little elephant's mouth.

"You must learn to use your trunk just like a *hand*," she says. "So you must bend your trunk, or turn it, or twist it, to get the thing you are holding exactly where you want it."

And that is the first great thing the little elephant has to learn— *how to use its trunk as [37] we use our hands.* After that everything else comes easy.

Now I am going to tell you about the childhood of the most wonderful elephant in the world, who actually lives to-day in the courtyard of a palace in India. He is the biggest elephant that ever was; that is why he lives in a grand palace, and does nothing except carry a King, or some other great man, on his back on days of festival.

In fact he was the leader among all the elephants in a long procession at a grand festival called the Durbar, held in honor of the King of England. On that day a lovely cloth of silk woven with gold was put on the elephant's back, and around his tusks were placed rings of solid gold studded with real diamonds, rubies, and pearls.

At another time he carried on his back the Crown Prince of Germany, when he visited India a few years ago; and at other times he has carried Grand Dukes of Russia and Arch Dukes of Austria when *they* visited India.

So you see, he is quite the grandest elephant in the world. He has a real name, just like a man, and it is written down in books with the names of all the grand officers of the palace. [38] His name is Salar Jung; so we shall call him Salar for short.

He was born in the jungle, and his Papa and Mamma were quite wild then. It was only after he grew up that Salar came to live in a palace.

Elephant Child "Swats" Tormenting Flies

But now about Salar's early boyhood. After his Mamma had taught him to swim, to eat from the boughs of trees, and to drink for himself by dipping his trunk into the water, she had another useful thing to teach him. In the jungle there are swarms of tormenting flies; they come buzzing around the elephants, and bother them, just as they bother us. Now, *we* can whisk off the flies with our hands, but how about an elephant?

Of course, you will say, his trunk is his hand; and so he can use the trunk to slap the flies or whisk them off. True, but the trunk will not reach more than halfway down the side of the body; and the elephant is too stiff to bend his body as we do; and his tail is too short to reach even a yard each way. Then how can he get rid of the flies where he cannot reach them? Just think! [39]

If he only could make his trunk *longer*! But how could he do that? Very simply! Of course he cannot actually make the trunk longer, but he breaks off a small bough of a tree and holds it at the end of his trunk; then he uses the bough like a fan, and whisks off, or brushes off, the flies with it.

And that is what Salar's Mamma taught him to do. After that he was very comfortable.

Not quite; he had just one more thing to learn from his Mamma, to make him quite comfortable. The sun gets very hot, and when the elephants are feeding from tree to tree, or marching through the jungle, they feel the hot sun on their backs dreadfully—although they have a thick skin.

Now, how could they guard themselves from the hot sun? Just think!

Why, just as *we* do, you will say, by using a kind of umbrella! Of course you mean that an elephant could break off a large bough, and hold it over his head and over his back! But his trunk would soon get tired of holding anything as big as that! Besides, he has to use his trunk all the time to feed! If *you* had only one hand, you could not eat with it and at the same time hold an umbrella over your head [40] with it! Then how *does* the elephant manage it?

Elephant Covers his Back from Hot Sun

I shall tell you. He breaks off many small boughs, one at a time, and lays them on his back with his trunk; he is careful to lay them in proper order, and to criss-cross them, so that the boughs will not fall off. In fact, he tries to arrange them very much like the thatched roof of a cottage. That is very clever of him, is it not?

But then he does something else, still more clever! When a cottager builds his thatched roof, he has to plaster the ceiling to prevent any rain or sunshine from creeping in through the little spaces between the thatches. So also the thatch on the elephant's back has many gaps, through which the hot sun can still beat down on his skin. So what does he do to fill up the gaps?

He cannot do anything to *plaster* his back; but I shall tell you what he does do. He just draws into his trunk a lot of dust from the ground; then he curls up the trunk over his back, and blows the dust over the gaps in the thatch on his back. Of course he has to do [41] that many times to fill up all the gaps; but at last, when he does not *feel* the sun any more, he knows that his back is quite covered.

Is not that a very wonderful thing for the elephants to think out, all by themselves? And that is what Salar's Mamma taught him to do.

But, a few years later, he came to the age when boys among men usually have to go to school. Then Salar passed to the care of his Papa. In feeding through the jungle, when all the elephants march and eat from tree to tree, Salar walked with his Papa, and began to learn lessons from him. And his Papa's way of teaching him was quite different from that of his Mamma, and often very funny!

How Elephants Walk under Water

The first thing he taught was at the stream at midnight. By this time Salar could swim quite well; so he was enjoying himself with the grown-ups. But his Papa kept watching him with the corner of his eye. Little by little he drew nearer and nearer to Salar, and waited till the youngster came to a part where the water was not at all deep. Then suddenly his Papa gave Salar a butt with his head. Down [42] went Salar under the water, snorting and spluttering and hollering.

"Hold up your trunk, you simp!" cried his Papa.

But Salar was too frightened to remember to hold up his trunk; so his Papa caught Salar's trunk in his own and hoisted it clear out of the water. Then what was Salar's joy and surprise to find that he could breathe quite well, though his feet were actually touching the bottom of the stream. Of course he kicked out, and tried to get up to the top of the water again. But—

"Stay there!" cried his Papa, giving him another butt, though still holding the youngster's trunk carefully out of the water.

Then Salar lost all fear of the water; he was not a bit afraid of being ducked, so long as the tip of his trunk was out of the water. So he learned to do a wonderful thing—he learned to remain completely under the water, so that his feet were actually resting on the bottom of the stream, with only the tip of his trunk out of the water. No other animal can do that.

And the most astonishing thing about it is that the elephants have taught themselves to do that trick; so that *a whole herd of elephants [43] can walk into a stream in time of danger, and disappear from sight*, the smaller ones standing in the shallow parts, and the full-grown ones standing in the deeper parts.

I have known of lots of hunters, who were chasing a herd of elephants and who saw the elephants run ahead toward a river, to find to their surprise, on reaching the river, that the whole herd had disappeared as if by magic. They saw nothing, and did not dream that the little things floating here and there, no bigger than your fist, could mean anything. But of course they were the tips of the trunks of the elephants hidden under the water.

To have thought out even that one trick for themselves proves that the elephants are the wisest of all animals, next after men folks. And they have thought out many more tricks, as I shall tell you very soon.

But now I shall tell you the next trick that Salar's father taught him. An elephant often has to break down trees in the jungle to clear a way for himself; or sometimes he has to do that to make an open space where he can lie down comfortably. So this is the way Salar's father taught him to break down trees. [44]

How Elephants Break Down or Pull Out Trees

First he chose a small tree, not much thicker than your wrist; this he pulled out easily with his trunk, just as you might use your hand to pull out a small shrub. Then he chose a tree about six inches thick. He tried it first carefully with his trunk; but the tree was too strong to pull out in that way.

So the old elephant put his foot on the side of the tree, and pressed with all his weight — as you see in the picture. The tree bent more and more, and then suddenly broke off near the ground with a loud crack.

"I can do that!" cried Salar, frisking around his father, impatient to show what *he* could do.

Salar looked around and saw a tree of about the same size. He made a dash at the tree, put his right foot on it, and —

His father winked, but said nothing. For all elephants love a joke.

Now the wily old elephant knew that this tree was a banana tree, although the fruit had not yet started growing on it. The tree looked quite hard and strong, but it was really very soft and easy to break, like all banana trees. But Salar did not know that yet! [45]

An Elephant Breaking a Tree with his Foot

[47]

Instead, when he pressed on it with his foot and put his whole weight on it, just as he had seen his father do to the other tree,—snap went the tree like a twig, and Salar tumbled head over heels and went rolling over the ground.

"Haw! Haw! Haw!" laughed the merry old elephant. "Did I not show you, silly, how to try it first carefully, with your trunk, before putting your foot on it?"

"Of course you did!" Salar said, remembering.

"That is what men folks mean when they say, 'You have put your foot into it.' You must remember *never to put your foot into anything before trying it first with your trunk,*" the old elephant went on to say. "Now watch me knock down a still bigger tree."

This tree was as thick as a man's body. After trying it first with his trunk and then with his foot, the wise old elephant put his back on it and *heaved.* Little by little the tree bent on that side, but not very much. The elephant stopped heaving, came around and looked at the tree. Then he began to heave from the *other* side of the tree.

You have seen a man trying to loosen a nail from a board? He first hits the nail on one [48] side, and then on the other side; and he goes on hitting the nail from side to side, till it is quite loose.

Well, that cunning old elephant did just the same thing to that tree; he first heaved the tree from one side, and then he heaved from the other side; and he went on heaving from side to side, till he loosened the tree from the ground. Then he pushed the tree with his foot, and it came out of the ground and fell with a loud thud.

And that is how Salar learned to heave with his body, though of course he could not loosen so big a tree just yet.

There were many other tricks that Salar learned from his father, and I shall tell you one of the best of them in the next chapter.

[49]

CHAPTER V

Elephants: The Tricky Trap

Salar and his father were going through the jungle, feeding from tree to tree, and from bush to bush. One day they saw a little clear space and in the middle of it a banana tree—just one tree. But beautiful bunches of ripe bananas were growing on it from a large stalk.

Salar just loved bananas. In fact, all elephants do, as they cannot get them in the jungle more than once in many months; for bananas grow mostly in plantations kept by men. So Salar ran toward the tree joyously.

But the wise old elephant had seen at once that the space all around the tree was rather level and clear of bushes. That was strange in the jungle, he thought!

Now, why did it look strange? Can *you* tell? Why was it strange that the space should be all flat and level, and clear of bushes? Just think!

Because in the jungle that was not natural! In the jungle the space should be all covered with grass and bushes, or at least with small shrubs of different sizes, just as you have seen in fields [50] which are allowed to grow wild. So somebody must have *made* the place level and flat, and cleared away the bushes! That is what the wise old elephant thought!

Then, also, he had seen that there was just *one* banana tree, with no other anywhere near it. That also seemed strange! Why? Because banana trees always grow in groups of many dozens, whether they are in the jungle or in a plantation.

"Halt!" the old elephant cried, just in time. Salar was not more than five or six yards from the tree when he heard his father's voice. I have told you before that, when an elephant child is told to do anything by his Mamma or Papa, he obeys *at once*, or else he might fall into some awful danger—just as a child in a town might get run over by an auto or a street car.

So as soon as Salar heard his father's voice, he halted just where he was. And that saved him, as you will see.

The Elephant Taps Suspicious Ground with his Trunk

His father came up to him, and looked around carefully. Then he *tapped on the ground with the end of his trunk.* [51]

"An elephant must always tap with his trunk when he is coming to suspicious ground, before he puts his foot on it," he said to Salar.

"What does suspicious ground mean?" Salar asked.

"Ground where there might be danger, though you do not *see* the danger," his father answered.

He went on a couple of yards, tapping the ground before him all the time. Then he suddenly stopped.

"Gr-r-r-rump!" he cried, "it sounds strange and hollow!"

Most carefully he put his foot forward and *felt* the ground with it, as an elephant always does when he thinks there is danger. Now the ground *bent down* a little just where he pressed it with his foot!

"I thought so!" he muttered.

Then he felt most carefully all along the *front* edge of the open space, first tapping it with his trunk, then pressing on it with his foot — of course with the toe end of the foot. And all along that front edge of the open space the ground bent down a little wherever he pressed it with his foot.

Then he came to the *right side* of the open space where the banana tree grew, and tried the [52] ground there also along the edge. And this ground too bent down a little wherever he pressed it with his foot.

He came to the *back* of the open space, and tried it in the same way. And there also the ground bent down a little along the edge, wherever he pressed it with his foot.

He came around at last to the *left side*, and tried that also. And there again the ground bent down in the same way.

"All four sides are suspicious!" he cried. "My son, this is *a most tricky trap*!"

And though he did not see them, a dozen men were hiding in the tops of trees all around. They were the hunters kept by a great Prince, who had ordered them to catch the big elephant and also the young one.

The hunters had first dug a huge pit. It was ten feet deep and twenty-five feet wide on each side; so it was as big as a large room. Then they had covered the top of the pit by laying many long bamboos right across from side to side and very close to each other; so it was just like the roof of a large room. And on the top of the bamboos they had spread a layer of earth—just like what you have seen in flower beds in a garden; and on that they had planted [53] grass, to make it look quite natural—only, they forgot that it might look natural for a garden, but not for a wild jungle. Or perhaps they thought that an elephant would not know any better!

And then they had gone to a plantation and fetched from there a banana tree, with a huge bunch of ripe bananas on it. They had set up the tree in the middle of that space; and as it would not keep straight, they had first driven a long bamboo rod right through into the ground, and then tied the banana tree to the top of the rod.

After doing all that, the hunters were hiding in the trees around. They were watching to see the big elephant and the little elephant come right up to the banana tree to eat the bananas, and get caught! For if any elephant stepped upon that place, the top would give way under his full weight, and he would fall right through into the pit.

But Salar's father grabbed him with his trunk, and pulled him away.

"Come away!" he said. "This is a most hideous trap!"

But Salar, who loved bananas quite as much as you love ice cream, began to cry. [54]

"I want the bananas; I want them; I do, I do!" he kept saying over and over again.

Now his Papa was very fond of Salar, but he did not know how to reach the bananas and not fall into the pit. He and Salar walked home slowly.

"I must think it over a bit," he said, scratching his head with a bough.

He came there the next day with Salar, and looked all around the place; but he could think of no safe way to get the bananas. The hunters also came there the next day, for by this time they were quite excited to see what the wily old elephant would do. In fact, it was from the chief hunter of that Prince that I heard afterwards what the elephant did do.

I must tell you here that these hunters had been watching the big elephant for many years, and trying to catch him by different kinds of traps; and that is how we know all about him and Salar. For when an elephant is very big and has fine tusks, people sometimes try for ten years to catch him, so that he may be used as the leading elephant of a grand palace.

Almost all the elephants you see in the zoo or in a circus were once quite wild in the jungle, and have been caught by some kind of trap. They [55] are then tamed, and finally trained to do tricks that men want them to do. I shall tell you all about that in another book, when you are a little older.

But now about Salar and his father. On the third day the big elephant came there again, with Salar; and again the hunters came and hid in the trees around. This time the big elephant looked farther into the jungle. Then he saw the long bamboos growing in a clump—the very clump from which the hunters had got the bamboos to make the trap. As the elephant looked at the clump of bamboos, a thought came slowly into his head.

He pulled out a long bamboo, and returned to the place where the trap was. He stood just outside the trap, and thought again for some time. Then he held one end of the bamboo in his trunk, pointed the other end to the banana tree just where the stalk of the bunch began, and gave a jab.

But he did not aim right, and the bamboo slipped off from the stalk. So he tried again, and gave another jab at the stalk. In this way, after trying many times, he managed at last to hit the stalk and break it. Down fell the bunch of bananas to the ground. [56]

Meanwhile Salar was jumping around his father for joy. But his father told him to keep still. He had not succeeded in getting those bananas yet! How could he get them out of the place of danger?

It puzzled him a long time. He poked at the bunch with the bamboo, but that only broke off one or two of the bananas. Then he poked at the stalk of the bunch, but the end of the bamboo slipped off it, as there was nothing on the bamboo to grip the stalk with.

So he drew back the bamboo and looked at that end of it, to see why it did not grip the stalk. Of course the end of the bamboo was all smooth, and could not grip anything at all.

Elephant Tricks the Tricky Trappers

Then at last another thought seemed to come into the wise old elephant's head. He put that end of the bamboo into his mouth and began to *chew* it; for an elephant has very strong teeth at the back of his mouth. As his mouth was very big, that clever elephant chewed as much of the end of the bamboo as his mouth would hold—and that was as long as your arm. So the end of the bamboo became like fibers, that is, like a bunch of hair, only very thick and rough. [57]

Elephant Pulling Bananas out of a Tricky Trap

[59]

Then that cunning elephant sat down on the ground and pushed the bamboo along the ground straight before him toward the bananas. When the hairy end of the bamboo reached the stalk of the bananas, he began to *twist* the other end of the bamboo with the tips of his trunk; for *an elephant can use the tips of his trunk in the same way that you use your fingers.*

He twisted and twisted many times, taking care to keep the hairy end of the bamboo pressed against the stalk of the bananas.

In this way the hairy end of the bamboo got knotted around the stalk. That was just what the wise old elephant wanted.

Then he pulled the bamboo slowly along the ground, as you see in the picture, taking care to give one or two more twists in case the knots came undone. He pulled the bamboo lap by lap; that is, he pulled the bamboo for about a yard, then he let go and took hold of the bamboo farther up; he pulled again for another yard, and so on. In this way he at last pulled the bunch of bananas quite out of the trap.

I need not tell you how he and Salar enjoyed that feast!

And the hunters, who were hiding in the trees [60] around, laughed and laughed at the trick the wily old elephant had played on them! For, as you see, he got the bananas and yet escaped from that trap! He beat the men at their own game!

But now I must tell you about other animals,—first about buffaloes. They are the brave knights who helped that timid little lady, the blue deer. They are just as wonderful as the elephants, in their own way.

[61]

CHAPTER VI

Buffaloes: The Knights of the Jungle

There are three or four kinds of buffaloes that live in different countries. The kind that lives in America you may have already heard about. I am sorry to say that hunters have killed so many of them, that there are very few buffaloes left in the United States now; and these few are kept in parks.

So in this book I shall tell you about another kind of buffalo, that lives in the jungles of India. These are the buffaloes that have to live in herds just because they have to guard themselves from the tiger. Yet they are much bigger than all other kinds of buffaloes in the world. Many of them are more than ten feet long, and a span taller than a tall man.

They have two huge horns which stand outward, one from each side of the head. Each horn is at least a yard long; and there are some buffaloes that have horns two yards long! (See the picture facing the next page.) [62]

So you can understand that this kind of buffalo is a strong and mighty animal. But still, if just one buffalo tried to fight a tiger, the tiger could kill him every time. Why?

Because the tiger is much *quicker* than the buffalo. The tiger could jump to the side to escape the buffalo's horns. Then the tiger could turn quickly, and strike the buffalo on the neck from behind. And though the buffalo's neck is very thick, the tiger himself is so strong that he could break the buffalo's neck at one blow.

So, to guard against the tiger, the buffaloes have to live together like a band of soldiers, so that the tiger never gets the chance of catching just one buffalo alone.

Now I shall tell you how these buffaloes live. They live in a part of the country where there is plenty of water, and lots of trees and grass. There is sure to be a stream or two in the jungles there, like the one where we saw the buffaloes drinking at midnight.

When the buffaloes are feeding in the jungle, and wandering here and there to find good grass to eat, they always try to remain somewhere near one of these streams.

Why do they do that? To drink, of course. [63]

The Buffalo that lives in India

[65]

And as the country is hot, they may want to drink more than once in the day.

Still, there is another reason why they like to be near water. Can you tell what it is?

"To bathe in the water, when it is hot," you may say.

That is quite true; the buffaloes do enjoy a good bath. In fact, they like to remain in the water for a long time, when the sun is very hot. Then they lie down in the shallow part, and remain neck deep in the water. And every now and again they dip their heads in the water to keep them cool.

But even when the sun is not at all hot, when the sky is cloudy, the buffaloes like to go into a stream or a pond. Why?

"Of course to wash themselves, and make themselves clean," you may say.

Buffaloes Cover Body with Mud against Flies

No, my dear, you are wrong this time! Like some little boys, buffaloes do not *want* to make themselves clean! In fact, the buffaloes go into the stream or the pond to *cover themselves with mud*! To *wallow*, as it is called. They do that by rolling in the mud where the water is shallow. [66]

And why do they want to cover themselves with mud? Because of the tormenting flies! Buffaloes of this kind do not have long hair on their necks, like the American buffaloes. In fact, they do not have much hair anywhere on their bodies—just like the ordinary cows which you have seen near your home. So they are very much tormented by the flies.

I have told you that an elephant can "swat" the flies with a bough which he holds in his trunk. But the buffalo has no trunk, and his tail can whisk off the flies for only a yard around. So, what can the buffalo do to guard other parts of his body from the flies?

The only thing he can do is to go down into the mud, roll about, and cover himself with the mud. Then he does not feel the flies at all, even if they swarm all over him. And he need not trouble to work his tail at all, as he is protected all over by the mud.

And when he comes out to feed again, if the sun happens to be very hot at that time, he does not mind it. Why? Because the mud on his body keeps off the sun. So, you see, the mud is useful to the buffalo in two ways.

But now come with me into the jungle while I show you all that the buffaloes do. [67]

You must *imagine* that I am taking you quietly through the jungle, where the buffalo herd is grazing right ahead of us. We are following them from behind. You must be careful not to make a sound. If you should tread on a rotten twig, the buffaloes would hear the sound as far away as a quarter of a mile.

In another book I shall tell you why all animals that keep their *ears close to the ground* while they are feeding can *hear a sound a long way off*.

But now let us hide behind this bush for a minute, and watch the herd. They are eating the grass as they walk along. But do you see the wonderful way in which they are arranged? It is just like the shape of the moon when it is new, that is, something like the letter C, and which we call a crescent.

You saw at the midnight pool that, when the buffaloes drink or march, they are in rows close together, like soldiers. But when they are eating grass, they could not be in rows; because then they would be too close together to pick out the best bits of grass. So, how could they have enough to eat, and yet guard themselves from danger? To do this they [68] thought of arranging themselves in the form of a crescent.

How Buffaloes Guard against Tiger while Feeding

It is a big crescent, as there are so many buffaloes that make it up. The ends of the crescent bend in toward each other, just as if the two tips of the letter C were to close up a little, leaving only a small opening between the tips.

The buffaloes have their faces toward the *outside* of the crescent. So, as we are following the buffaloes from behind, we are looking at them through the gap between the tips.

There are only bulls in the line making up the crescent; the cows and the calves come behind them, so that they are *inside* the crescent. So you see, while the buffaloes are grazing and moving along, if they meet any danger, the Papa buffaloes will face the danger. And as the Mammas and the children are inside the crescent, they are quite safe.

This is the way the buffaloes feed and move along:

The Papas on the outside of the crescent tear off a mouthful of grass, with one or two [69] bites, and walk on a step or two while they are munching the mouthful. Then, with another bite or two, they take a fresh mouthful and walk on a step or two while they are munching that. In this way they leave enough grass for the Mammas and the young buffaloes that are following them.

But now let us come out of this thicket, and go after the herd very quietly from behind. We shall see some wonderful things.

You notice at once that the Mammas and the children do not show any fear at all, as they graze along; they are quite sure that the Papas are taking good care of them all the time. The little ones even play about here and there.

See that very young calf! He is playing about near the middle of the space. He is only a few weeks old, and not much bigger than the calf of the ordinary cow. Watch and see how playful he is! He is just like any other calf. His Mamma is grazing along quietly, and he is now standing still for a minute, looking at nothing. A calf and a baby can do that quite well—just stare, and yet look at nothing.

But now this buffalo calf rushes to his Mamma [70] very suddenly, and has a mouthful of milk. He does not seem to want more than a mouthful at a time. So he looks up suddenly, and stares. Then just as suddenly he plunges into a frantic race over the ground, all by himself.

The race also ends suddenly—after going only ten yards. Then he stops there for a minute, stares, and trots back to his Mamma for another mouthful of milk. After that he looks up again for a minute, stares at nothing, and plunges into another mad gallop all by himself.

So you see that he spends his time doing two things—having a mouthful of milk, and then a mad gallop. And he does both very suddenly. He likes to have his joys suddenly.

A kitten or a puppy dog is different, and is nearly always doing something. It tumbles head over heels, or chases its own tail, or keeps frisking about in some way or other most of the time. But the buffalo calf is not like this; and when you see him standing quite still, staring at nothing, you can never tell whether he is going to be hungry for a mouthful of milk the next minute, or whether he is going to break into a frantic race.

But, you may ask, while he and all the [71] other calves are playing about like that, is there no danger?

No, there is no danger, for the Papas are taking good care of the Mammas and the children, as I have told you before.

But, you may say, the Papas do not seem to be doing anything; they are just feeding and moving along. Then how are they taking care of the Mammas and the children?

Yes, but look carefully! See how close the horn of one Papa is to the horn of the next one! Why, there is not more than a couple of yards between the two! If there were any sudden danger, it would not take more than two or three steps for them to close up, and stand horn to horn.

How Buffaloes Know Danger is Coming—Three Ways

"But how could they *know* if any danger were coming?" you may still ask.

They could know it in three ways: they could *smell* the danger, or *hear* it, or *see* it. I shall tell you how they do all that.

First, if the danger came from the direction in which the wind was blowing, they would sniff the air, and so *smell* the danger. If the danger were [72] a tiger, the buffaloes could smell him half a mile off; that is about as far as ten blocks in a city. And if the wind were not blowing that way, the buffaloes could still smell the tiger five blocks away. They could smell the tiger, or any other danger, even if it came from behind.

The second way of finding out the danger is to *hear* it. As I said a little while ago, if you should put your foot on a rotten twig, the buffaloes could hear the sound of it as far off as five blocks. And even if the danger came from behind, or from the side, or from anywhere, they could still hear it coming, if it made the least bit of sound that you and I could not hear.

The third way of finding out the danger is to *see* it. The buffaloes do that by keeping a lookout nearly all the time. I shall show you how.

Just watch for a minute the buffalo in the middle of the crescent; he is the leader of the herd. We can see him only from the back; but as he is the biggest and tallest buffalo there, we can make him out quite easily. He is grazing quietly, and then moving along.

But see, what is he doing now? Why, he [73] is looking up, straight ahead of him! No, he sees no danger there. So he gives a glance to his right, and then to his left. No, there is no danger there either. So he puts down his head, and starts feeding again.

Thus, you see, every now and again he looks to see that no danger is coming from anywhere in *front* of the herd.

But what if any danger came from the *side* of the herd,—right near the end of the crescent,—or even from the *back* of the herd?

Buffalo Sentinels

Let us see what the two buffaloes at the two ends of the crescent are doing. They are the watchers, or *sentinels*, as they are sometimes called. They keep a lookout nearly all the time.

Do you see the one on our left? After every two or three mouthfuls he stops, and takes a look around; he even looks right to the back. Then he takes four or five strides to catch up with the herd, and starts grazing again. Then in a minute or two he takes another look around in the same way.

And the sentinel on our right is doing just the same. Yes, the herd is quite safe; the two [74] sentinels are sure to see if any danger comes from their side or from the back.

"But will not the sentinels have less to eat, if they are watching half the time?" you may ask.

Yes, that is quite true. So all the Papa buffaloes take turns being sentinels. After a while the two sentinels from the ends move up toward the middle, and the next ones then begin to keep watch. And they keep changing places like that from day to day. That makes it quite fair for everybody.

When they go to sleep also they are arranged in the form of a crescent; but the two ends are closed up, so that the Papas make a ring, while the Mammas and the children sleep inside the ring.

When the Papas lie down, they are closer together than when they are feeding; and they still keep their heads pointed to the outside of the ring, so that they can get up in a minute, and be quite ready to drive off any tiger. Of course they have sentinels keeping watch all the time.

But now let us see other wonderful things that the buffaloes do, while they are feeding.

We must be very careful how we follow the [75] herd. The ground is now changing, and getting quite rough; so the grass is getting scarce here and there. The buffaloes have not enough grass all the way; so they have to walk on a few yards without eating, till they come to the next patch.

Some of the buffaloes are even having a bite at fresh young shrubs in passing, as they will eat anything green, when they have not enough grass.

But see! The buffaloes are spreading out, as there are not even enough shrubs in one place. You can see gaps in the line of buffaloes now. And the gaps are getting bigger and bigger! Let us watch a few minutes.

Now the gaps are very wide, as some of the buffaloes are lagging behind; and some are turning too much to the side in trying to reach a mouthful from a shrub or a bush here and there.

Why, what is happening now? Some of the buffaloes cannot even see one another now, because of the bushes between them! What are the sentinels doing? And what is the leader doing? Suppose a tiger suddenly comes—

But do you *hear* that? [76]

"Moo! Moo!"

That is the leader. He has just found out that the herd is spread out too far; so he is calling. He is saying, "Where are you?"

"Moo! Moo!" Do you hear that answer? It comes from the sentinel on the right, who is very far away now; but still he has heard the call. His answer means, "Here I am!"

And "Here I am!" comes the answer also from the sentinel on the left.

"Then close up!" cries the leader.

Each sentinel moves up toward the place from which he heard the leader's voice come. And on his way there he tells all the buffaloes he meets to move up also. Besides, all the buffaloes hear the leader's voice too; so they begin to close up at once.

Is not that a wonderful way of bringing up all those that are lagging behind?

But let us watch the herd again. They have closed up now, and there is no big gap in their line. The ground is level again.

Let us move on from thicket to thicket, and come as near the buffaloes as we can.

What is that? See! The sentinel on the right is looking hard at that jungle grass far away to the side. This kind of jungle grass [77] grows very tall, taller than a man. But why is the sentinel staring at the tall grass? What does he see there?

Yes, there, far away, something is happening! The jungle grass is waving gently, but just in one place! What is making the tall grass wave like that? Is it the wind? No, it cannot be the wind! Why not? Because if it were the wind, *all* the grass there would wave. Then what is making the tall grass wave in just one place?

It can be only one thing! Some *animal* is hiding there in the tall grass! And as the animal is coming nearer and nearer to the buffaloes, he is making the grass wave!

See, the sentinel has guessed that also! What is he doing now? Can you *hear* him? He gives a bellow, deep and long.

"Danger! Look out!" That is what he means.

The whole herd hears him. They all close up as near together as they can!

Quick! Let us get up on that tree near by! *We* are in danger as well as the buffaloes!

One branch higher—and another! Now we are quite safe! But see what the buffaloes are doing! [78]

Buffaloes Make a Ring when Tiger Comes

The two ends of the crescent have come close together, and all the Papa buffaloes have made a perfect ring around the Mammas and the children. The Papas are facing the outside of the ring; so they can meet the danger from whatever side it comes.

Why do they do that? Look again at the grass! The tall grass is waving nearer and nearer. So, the animal that is in the grass is coming nearer and nearer.

He comes right to the end of the tall grass. There he makes a gap in the grass, and walks out into the open. It is a tiger!

He was trying to sneak up to the buffaloes; but the sentinel found that out. And now the bull buffaloes are ready for him. The tiger growls in rage. He prowls round and round the ring of bull buffaloes, as you see in the picture. But he dare not try to break through those horns.

He roars with fury, shaking the ground; it is just like thunder. The jungle around is taking fright at the roar. See! All the small animals rush out in fright—wild pigs, wild goats, and all sorts of small deer. [79]

The Tiger and the Ring of Buffaloes

[81]

In their fright they run hither and thither very stupidly. That is exactly why the tiger roars—he wants to make the small animals behave so stupidly, in their fright, that some of them may make a mistake and run straight into his jaws.

See! The small animals scatter to right and left, trying to reach a bush or thicket. But some are cut off from safety, as the tiger stands in their way. What can they do?

Small Animals Find Safety in Buffalo Ring

Yes, there is the ring of buffaloes! So those small animals rush straight toward the ring and creep inside—and the buffaloes raise their heads to make a way for them under the horns. Some of them, like the wild goats, jump *over* the buffaloes' horns to get inside the ring. Anyway, the small animals reach safety inside with the Mammas and the children of the buffaloes.

The tiger stands outside the ring, and still roars in fury. But now nobody is afraid. The bull buffaloes paw the ground impatiently with their hoofs, and rattle their horns. They are going to charge!

But that tiger does not wait for the charge. [82] He does not want to be trampled into a mess. So he slouches away, growling and snarling.

So, as you see, the bull buffaloes guard the Mammas and the children from danger, and they also guard all small and weak animals that come to them for safety.

Did I not tell you that the buffaloes are the Knights of the Jungle?

Tame Water Buffaloes Plowing in the Rice Fields

[83]

CHAPTER VII

Taming the Buffalo

Buffaloes do not always remain wild and wander about in the jungle. Men need buffaloes. Farmers want to use them for plowing the ground, in the same way that farmers in America use horses for plowing.

This kind of buffalo also lives in Italy, and because they are so fond of water they are called *water buffaloes* there. But in Italy they are not wild any more, as they have been tamed and used by men for a long time.

I shall tell you how the men catch the buffaloes from the jungle in India, where they are still wild.

They catch the buffaloes in many ways. The easiest way is to find some stream or pond where the buffaloes are fond of going. Then the men take strong nets made of ropes, and spread the nets under the water. So when the buffaloes come to bathe or roll in the mud, some of them are caught in the nets.

Then the men rush in from their hiding place and drag out the nets. Of course, those [84] buffaloes which are not caught run away. But those that are caught struggle fiercely. After a time they get tired of struggling, as the nets are too strong for them to break.

When the buffaloes have become very weak from struggling, a lot of men rush up and tie a stout rope around the neck of each buffalo. The rope has two ends, one on each side of the buffalo, and each end is quite long.

A dozen men haul at the rope, and the buffalo has to get up and march with them. In this way the men bring the buffaloes one by one to the village.

How do the men tame the buffaloes? That is quite easy, if they already have a few tame buffaloes which they may have caught and tamed some time before. And as the people have been doing this for many, many years, they always have some tame buffaloes. So this is the way the men treat the wild buffaloes:

Wild Buffaloes Tamed Quickly by Kindness

They put the wild buffaloes and the tame ones together in a pen, or corral. Inside the corral there is a pond. In the deep part of the pond there is plenty of good water to drink; and in the shallow part of the pond there is [85] plenty of mud in which the buffaloes may roll about and wallow.

The men keep the buffaloes there together for many days, the wild ones and the tame ones. Every day the men throw into the corral plenty of fresh grass, which the buffaloes can eat all day.

Now, what more could the wild buffaloes want? They could not be treated any better! They have plenty to eat, plenty to drink, and plenty of mud in which to wallow. The tame buffaloes soon make friends with them, and talk to them in their own language.

"You will not be any better off in the jungle," the tame ones say to the wild ones. "Here you do not have to walk about all day to get enough to eat, and then walk a good way to find water to drink, or a place in which to wallow. And, also, we have no fear of tigers here. What more do you want?"

So in a few weeks the wild ones become quite tame. Still, even after that, the old and the new ones are always kept together, and soon they become like one herd.

Afterwards, when the farmers use them for plowing, they always hitch to the plow one buffalo that has been tame for a long time, [86] and one that is newly-tamed. Then it becomes easy for the new one to learn the work by just doing as his friend does.

The farmer uses the buffaloes for plowing for only a few hours, and he gives them plenty of time for wallowing and enjoying themselves. So, even if they have to do a little work, the new buffaloes soon see that they are really much better off living in the village than running wild in the jungle.

After the plowing season is over, the buffaloes have no work at all. They can wallow all day, if they want to.

When all the new buffaloes are quite tame, they are not kept in the corral any more, as they would never think of running away now. They are allowed to lie about and sleep in a little plot of ground somewhere in the village. By daytime they are taken out into the fields outside the village, and allowed to graze as they please; and as there is always a stream or a pond near, the buffaloes can go into the water or the mud whenever they like.

So, as you understand, the buffaloes very soon become quite tame. Why? Because they are treated kindly. Please remember that. *Most wild animals can be tamed if treated kindly.* [87]

Now I am coming to the nicest part about the buffaloes. It is the nicest part because it shows how the buffaloes can even be made to love us.

I have just told you that the buffaloes are taken out into the fields to graze. Well, then, somebody has to do that in the morning, and somebody has to bring them home in the evening.

Can you tell who does that? Why, there is a herdsman to do it, you may say. Quite true. But the herdsman does not bother to do a simple thing like that every day.

Little Boys Take Charge of Buffaloes

Then who does it? I shall tell you. The little boys of the village! They are about five or six years of age. They are not old enough to go to school, and not old enough to do any work; so they can play all day.

The most useful thing they can do is to take charge of the buffaloes. The boys soon learn all the buffalo calls—"Come out to graze," "Come to wallow," or "Come home now." And the wonderful thing is that these huge animals soon learn to obey these calls. When the boys call to them, the buffaloes do just as they are ordered. [88]

The buffaloes soon learn to love the little boys. You know how fond of us an animal can become—especially a dog or a horse. Still, I do not think that any animal can show such love for us as the huge buffaloes do for the little boys who act as their herdsmen.

Why? Because the little boys *share the same mud* with the buffaloes! Boys and buffaloes mix very well with mud! The little boys tumble about in the mud on the side of the bank where the buffaloes may be wallowing. Or the boys will splash about in the water where the buffaloes are lying neck deep to keep cool. Or they will climb up on the buffaloes' backs for a while, then tumble off and play again.

Even when the buffaloes are grazing in the field, the boys may be near them, playing hide and seek, and running in and out between the buffaloes' legs, or under their horns. So the boys are with the buffaloes all day long.

How the Big Buffaloes Love the Little Boys

It is quite wonderful to see a little boy actually twisting a huge buffalo's tail. As I have told you, a buffalo is often more than ten feet long, and taller than a tall man; and it has horns that [89] reach out more than a yard from each side of the head. This huge animal could charge and smash up a big wagon as easily as if it were a match box; and yet he will stand still and let his tail be twisted by any little tot in the village.

Sometimes you may see a sight like this: A huge buffalo is grazing hungrily, and a little boy comes up and stands right in front of him.

"Put up your head!" says the boy. But the buffalo goes on feeding hungrily.

"Put up your head, or I will spank you!" says the boy. But the buffalo still goes on feeding hungrily.

Then that tot raises his small hand and spanks the huge buffalo on the jaw. The buffalo puts up his head, and rubs his nose lovingly against the boy.

Well, why not? You have seen a baby pulling his Papa's hair. The Papa just loves the baby all the more for it. So it is with the buffalo and the little tot. And it would not matter a bit whether the tot were a little boy or a little girl. The big buffalo is fond of both.

And now I shall tell you a wonderful true story about a buffalo and a boy.

[90]

CHAPTER VIII

The Buffalo and the Boy

In a village there were many tame buffaloes, and among them thirty bull buffaloes. The little boys of the village took charge of them every day. The smartest boy among them was called Gulab. He was six years of age.

Gulab knew quite well each of the thirty bull buffaloes, and was a friend of each. Sometimes he alone had charge of them, and took them out to graze and to wallow. That was because his father was the herdsman.

The buffaloes loved Gulab, and they did exactly as he told them to do. When he was going to take them to the fields, he would just stamp his little bare foot and call out to them "Stand in rows!" And the huge animals would stand in rows, one line behind another.

Then Gulab would come around to the side, and see if each line was straight. If the line was not quite straight, and a buffalo happened to be standing too much this way or that, Gulab would walk up to the buffalo and spank [91] him on the jaw. Then the buffalo would move into line, exactly as Gulab wanted him to do. Or, if a buffalo happened to be standing too far behind, Gulab would come around to the back and twist the buffalo's tail, and the buffalo would move up into line.

Then, when the whole herd was in the right order, Gulab would come to the front of the herd, and walk up to the biggest bull.

"Bend down your head, Baldo!" he would order.

And Baldo, the biggest bull, in the middle of the front line, would bend down his head, and Gulab would climb up by one of the horns, scramble up Baldo's neck, and sit down on his back.

"March!" Gulab would order—and the whole herd would march.

Now, a few miles away there was a grand palace. In the palace was a little Prince, whose father was a Rajah—that is, a kind of king.

The little Prince's birthday was coming, and his father ordered grand feasts for many days.

The Rajah had six English friends, who were quite big men. The Englishmen were very fond of tiger hunting, so the Rajah wanted to order a tiger hunt for them. But it is not easy [92] to have a tiger hunt just when you want to have it. Why not? Because the tiger will *not* come out and be hunted just when you want him to. He would rather stay in his den.

So for a few days no one heard of a tiger prowling about. Then suddenly a strange piece of news came from that village where Gulab lived. It happened in this way:

One day Gulab took out the buffaloes to graze and to wallow. The buffaloes lay down in the shallow water for a while, and Gulab splashed about or tumbled in the mud near them. Then he got tired of doing that, and came out on the bank and played about there for a while.

Suddenly he heard a strange sound. It was one of the buffaloes, who had stood up in the water and was giving a low, deep bellow. Two or three other buffaloes stood up also, and gave a low, deep bellow. Then all at once the whole lot of them began to come out of the water.

Gulab stopped in his play to see what was wrong. But he could see nothing.

"What's the matter, Baldo?" he asked. "What's wrong, Chando?"

But the two biggest bulls scrambled up the bank, and came rushing toward the boy. All [93] the other bulls came also, and some went past him on the right side, and some went past him on the left side. Then suddenly Gulab knew what it all meant!

A snarl—a growl—a roar, he heard. A flash of yellow leaped out of the jungle, and came toward him with a huge jump. It was a tiger!

But already the buffaloes were making a ring around Gulab. Then he knew what had happened. The tiger had seen him from the jungle beyond, and had been trying to creep up to him quietly from thicket to thicket. But the buffaloes had *smelled* the tiger in time, and had run out of the pond to save Gulab. And now they had made a ring around him.

Gulab stood in the ring and looked with large round eyes, for he was more frightened than he had ever been in his life. He was only a little boy, and had never seen a tiger face to face.

The tiger growled and snarled and roared. Then it came round and round the ring, trying to find a gap between the horns to get at the boy. But there was no gap between the horns.

Then little by little the fear left Gulab's heart. Something inside him told him to be brave. He walked up to Baldo. [94]

"Baldo, let me up!" Gulab said to him, standing behind the buffalo. And Baldo lowered his body behind, and bent his hind legs at the knees.

Gulab took hold of Baldo's tail in both hands, and put his foot on Baldo's hind knee, which was now bent quite low. In that way Gulab climbed up to the buffalo's back, and sat on it, holding on to Baldo's shoulders.

Then, being quite safe on the buffalo's back, Gulab glanced around and called to the buffaloes at the back of the ring, "Open out!" And the buffaloes opened out at the back of the ring, and made a crescent. Then they moved still farther around, and the crescent became one long line, facing the tiger.

Gulab gave one glance to right and left, to see that all were ready. Then—

"Charge, brothers, charge!" he cried to the buffaloes.

Then his big brothers, the buffaloes, charged with thundering hoofs and fiery nostrils. The tiger gave a huge leap to the side to get away; but the buffaloes on that side opened out and headed off the tiger. On to the front again the tiger was forced to turn—and run for his life before the furious herd. [95]

The buffaloes chased and chased that tiger, across field and jungle, over hedges and ditches, through brambles and bushes and thickets, till at last the tiger jumped across a ravine and ran away growling and howling and snarling, like a low thief who is chased out of a village at night.

The ravine was a deep hollow in the ground, like a huge ditch; and it ran all the way across the ground; so the buffaloes could not get over it, as they cannot jump as far as a tiger. Then the buffaloes returned to the village, and Gulab gave the news about the tiger.

Some of the village people ran to the palace, and said that the tiger might be still hiding somewhere on the other side of the ravine. So the six Englishmen went around to that side to hunt the tiger. They found him and wounded him four or five times. But it takes a lot more than that to kill a tiger. The tiger ran out, got past the hunters, and came back again across the ravine. Here he hid in a dense thicket, and would not come out and be hunted to please anybody.

Now, when a tiger is hiding in a thicket and will not come out and be hunted, there is only one way to *make* him come out. [96]

What is that way? Can you tell?

Why, of course, the bull buffaloes!

So the herdsman brought up the thirty bull buffaloes, and drew them up in a long line in front of the thicket. And on the other side of the thicket the six Englishmen got up into trees, and pointed their guns at the thicket.

Then the herdsman ordered the buffaloes to charge, and they charged right through the thicket, trampling it down and cutting it up into lanes; so the tiger *had* to run out on the other side. But on that side the six Englishmen were waiting for him; and they all fired at the tiger at once, and all hit him. They used a kind of bullet that broke up into a hundred pieces right inside the tiger.

But the tiger still kicked and kicked, and would not agree to be dead at once, as any other animal would. People say that a cat has nine lives; then a tiger must have ninety-nine lives. So this tiger jumped about, torn up as he was, and glared at the Englishmen in the trees, trying to get at them, while they were loading their guns for another shot.

But the buffaloes went on charging, and caught up with the tiger. They rushed upon him, and now the torn-up tiger could not get [97] away. So the buffaloes trampled upon him, and then the tiger agreed to lie still and be dead, really and truly.

The six Englishmen began to climb down from the trees, as they thought the excitement was all over. But the herdsman called out to them at once:

"Please go up again—quick! Don't let my buffaloes see you!"

For I must tell you now that buffaloes do not like strangers. They may be very fond of their own friends in the village; but if they should see a stranger, they would charge him just as quickly as they would charge a tiger. And the Englishmen would look quite strange to the buffaloes.

So the Englishmen remembered that, and stayed up in the trees till the buffaloes were taken away.

The buffaloes were taken to the pond; and as the herdsman would not bother to stay with them there, he left the buffaloes in the pond to do as they pleased till evening.

The six Englishmen had their lunch there, when they got down from the trees. They gave their guns to their servants, to carry away to the palace. Afterwards the Englishmen [98] walked about, smoking their cigars, as they did not want to return to the palace so soon.

But four or five hours passed, and still they had not come back to the palace. It was nearly evening, and still they had not come.

And in the village Gulab said to his Papa, who was the herdsman, "Papa, I shall bring the buffaloes home now."

He went to the pond. But the buffaloes were not there! He shouted, whistled, and gave all the buffalo calls he knew. But no answer!

He looked about, and searched everywhere, but he could not see the buffaloes. What had become of them?

Then he happened to look far to the side, toward a lot of tall trees. Something was happening under the trees! He could see a lot of things moving there, but he was too far away to see what they were.

He ran toward the trees. Yes, they were the buffaloes! But why were they there? And why were they behaving like that?

For he saw that they were pawing the ground angrily, and tossing their heads and rattling their horns. And what was very strange, the buffaloes were not looking at anything on the ground in front of them. They were looking *up*, at the trees!

Then Gulab glanced up into the trees, and saw at once why the buffaloes were behaving like that. But he did not waste a minute. He ran to the buffaloes, shouting:

"Down, Baldo! Down, Chando!"

But the two biggest bulls and all the others glared at the trees and snorted in fury.

"Down!" Gulab shrieked. "Down, or I shall spank you!"

He rushed to Baldo, and spanked him on the jaw. He rushed to Chando, and spanked him on the jaw. He rushed from buffalo to buffalo, and spanked each one on the jaw.

Then the huge animals that had charged the raging tiger, and that were now fierce themselves, obeyed the little boy. They blinked, then one by one lowered their heads. Gulab climbed up by Baldo's horns, and seated himself on his back.

"Now turn around, all!" he ordered. And the buffaloes slowly turned away from the trees.

Gulab looked back over his shoulder, and said to the six Englishmen who were up in the trees: "You may come down now. My buffaloes won't hurt you a bit, because if they try to I will spank them!"

Then the little boy took away the buffaloes, and the six big Englishmen came down from the trees quite safely.

And now, do you understand what had happened? I shall tell you. The Englishmen had forgotten what the herdsman had told them—about keeping away from the buffaloes. The Englishmen had walked about, and had finally come near the pond where the buffaloes were.

Then the buffaloes had come out and charged them. The Englishmen had run and run, and had just managed to reach the trees. But the buffaloes had come there after them! So the big Englishmen had to stay up in the trees, and wait for some little village boy to come and take away the furious buffaloes.

I have told you this story, my dear (and it is a true story) just to show you what kind of an animal the buffalo is—at least, this sort of buffalo. Even when he is furious, he will do anything for the little boy whom he loves.

But as it is a true story, I must tell you one more thing that happened—and I am sure you will be delighted to hear about it. The six Englishmen went to the palace, and laughed [101] and laughed, and told all about it to the little Prince whose birthday it was.

Then the Rajah, who was the little Prince's father, said that Baldo and Chando should not be made to plow any more, or do another bit of work in their lives. Why? Because Baldo and Chando had first helped to save Gulab from the tiger at the pond, and then afterwards they had helped to hunt the tiger.

So after that, Baldo and Chando were allowed to walk about the village as they pleased, and nibble at anybody's hay or grass, and splash in anybody's pond, and wallow in anybody's ditch, rut, or mire.

And what was little Gulab's reward for saving the six Englishmen? Well, the little Prince, whose birthday it was, came and took Gulab by the hand, and brought him to the grand palace, and gave him lots and lots to eat—cakes and ice cream and candy—so that Gulab went home that night very full and very happy.

[102]

CHAPTER IX

Deer and Antelope

The buffalo has many relatives among other animals which also have *horns*. In fact, all animals that have horns are some relation to each other—first cousin, second cousin, third cousin, and so on.

The buffalo's nearest relatives are the ordinary cows and bulls that you see in the fields.

"But the sheep and the goat also have horns," you may say. "Are they also cousins?"

Yes, they are. In the same way the *deer* and the *antelope* are also cousins to each other. I am now going to tell you about them.

The deer and the antelope are not exactly the same kind of animal, as you might perhaps think. As I said, they are only cousins. If you look at them carefully in the pictures on pages 103 and 109 you will see which is the antelope and which is the deer—just as you can tell a sheep from a goat.

[103]

Antelope
Photograph of a group in the American Museum of Natural History, New York

[105]

First see the picture on page 103. These are *antelope*. Look at the horns carefully. They are something like a cow's horns; only, a cow's horns are sometimes bent and twisted in different ways. But the antelope's horns point upward, and are much longer than a cow's horns. They sometimes look almost like a pair of long and thick spikes, pointed at the top.

Now look at the picture on page 109. These are *deer*. Look at the horns carefully—only, they are not called horns when the animal is a deer, but *antlers*, which is a special name. So take a good look at the deer's antlers. There are two of them, and they grow from the top of his head, like the antelope's horns.

But look again. The antlers *start* from the head as *two* spikes, but higher up each antler branches out into *many* parts. In fact, near the top each antler looks something like the branches of a small tree without leaves.

So now you can always tell which is an antelope and which is a deer: the antelope's horns have no branches, but the deer's antlers have many branches.

Horns and Antlers Different in Three Ways

The antelope's horns and the deer's antlers are also different in other ways, which you [106] cannot see in the pictures. So I shall tell you about them:

1. The antelope's horns are *hollow* inside, and made of the same kind of thing as the *hoofs* or *nails* of an animal, only they are thicker and harder. But a deer's antlers are *solid*, and made of *bone*.

2. Both the Papas and the Mammas among antelopes have horns. But among most kinds of deer, only the Papas have the antlers; the Mammas have none.

3. Among antelopes, when once the Papas and the Mammas have grown their horns, they keep them always. But among deer, the Papas throw away their antlers every year, and grow *new ones*. That seems very wonderful! I shall tell you more about it soon.

But now I shall tell you, little by little, all the wonderful things the deer and the antelope can do. I shall begin with the deer, as there are many kinds of deer in America.

Of course, in America there are not such wild jungles as in countries which are hot all the year round. Still, there are many places in the West and a few other parts of America where there is some kind of jungle and plenty of forest. A forest is a kind of jungle, only it has more [107] trees, and fewer thickets; but wild animals can live there just the same.

Elk and Other American Deer

The biggest kind of deer in America is the *moose*; in fact, it is the biggest kind of deer in the world. The second biggest is the *elk*; he is nearly as big as the moose. Some people think that the moose and the elk are exactly the same kind of deer, but that is not quite correct. In this book I must not make it too hard for you to understand, by telling you how they are different. So I shall tell you all about the elk, as his picture is on page 109.

Once upon a time elks lived in all parts of America, but now they have been killed off by hunters in most parts, and are found wild only in the Far West.

The elk is a fine fellow. At the shoulder he is as tall as a man, and is as heavy as six men. He lives in places where there is plenty of forest—that is, plenty of trees. Why trees? Because he needs them in winter—for then the bark is his food!

In summer he has plenty to eat—leaves, twigs, and grass. But when the winter comes, and the leaves fall, and the ground is covered [108] with snow, the poor elk would starve and die, if he did not have at least the bark of trees to eat. And very little bark he gets for many days at a time.

Here I must tell you that some kinds of deer are among the most *hardy animals*; that means that at times they can live on very little. There is a kind of deer, called the *reindeer*, that lives in the frozen North, where there is snow and ice almost all the year round; and the reindeer has nothing more to eat for many days than a little bit of moss or seaweed.

But there is another animal, not a deer, that is still more hardy: he can go a whole week without eating or drinking—and do work all the time! That seems very wonderful. But I shall tell you about that animal in another chapter.

Now about the elk. His antlers are fine! You can see in the picture how huge they are. And yet, would you believe it, he grew them in only five months! I told you a little while ago that a deer throws off his antlers every year, and grows new ones. I shall now tell you how the elk does that.

[109]

Elk
Photograph of a group in the American Museum of Natural History, New York

[111]

In the middle of the winter, the elk's antlers break off bit by bit. In a few weeks they have all fallen off, leaving the elk's head bare, with just a ridge or rough stump on it. Then, early in the spring the new antlers start growing from the top of the stump. They grow very fast, and in five months are as huge as ever.

But while the new antlers are growing, they are not hard. As yet they are soft and tender, and all that time they have an outside covering like hairy leather, to guard them from harm. But as soon as the elk feels that his antlers are quite grown, and are strong and hard, he strips off the outside covering by rubbing the antlers against trees.

Of course, while his antlers are still growing, and are soft and tender, the elk cannot use them to fight another animal; so during that time he hides in the bushes. But as soon as his new antlers have become hard and strong, he is very brave again, and is ready to fight!

Does the elk fight much? He does! He fights most awfully when he has his new antlers. What he fights about, and with whom he fights, I shall tell you in another book.

There are a few other kinds of deer in America, but the funniest of them is called the *mule deer*, which lives along the Rocky Mountains. He is [112] called the mule deer because he has very long ears, like a mule's ears. And perhaps you have seen a mule bucking—that is, jumping about while holding his legs quite stiff. Well, the mule deer can buck just like that.

And while he is running at a gallop, he will often jump off the ground with stiff legs, and then hop on and on many times like that, with stiff legs, finishing up with another gallop.

That makes him look very funny, and because he jumps like that people in Canada sometimes call him the *jumping deer*.

Other Kinds of Deer

I must now tell you about some other kinds of deer that live in jungles and forests in other countries.

The *fallow deer* lives in Europe. When he is wild, he lives in a forest; but when he is tame, he lives in a park. He is a small deer, about the size of a donkey. His coat is very soft and glossy and beautiful. In winter his coat looks dark brown, and his legs and the under part of his body are light brown. But in summer his coat becomes a lovely light red in color, with white spots jotted all over it. Then he is very handsome. [113]

In India also there is a handsome deer, which the people call the *lion deer*. He looks quite gentle and mild. Then why do the people call him the lion deer? Because he has a lovely coat, golden yellow in color. You could see him far across the open field, if he only stood there. But he is so timid that he does not often come out in the open.

And why has he a yellow coat? Because he lives in a place where there is plenty of yellow grass; and if he stood right in the middle of the grass, and did not move, nobody could see him. Even if a tiger were looking for him, and the deer stood quite still in the grass, the tiger could not find him.

In another chapter I shall tell you how other animals have on their bodies the *color of the place where they live, or where they want to hide*.

Barking Deer—One of the Wonders of Nature

Now I am coming to one of the nicest kinds of deer in the world, and I am sure you will just love him! He lives in India, and is called the *barking deer*; only, he is not exactly a deer, but an antelope. You remember what I have told you before, about an antelope having a different kind of horns? Still we must call him [114] the barking deer, as people have already given him that name.

He is very small, about the size of a goat. If there is any danger from an enemy, the barking deer is small enough to hide in any little bush or behind a fallen tree or log; or else he can run away very quietly through the under bushes. And he runs so quickly that his enemy soon loses sight of him.

He is called the barking deer, because he can bark or yap almost like a dog. But, you may ask, why does he want to bark at all, if he is afraid of some enemy? Will not the enemy hear him, and then catch him?

Yes, that is quite true. And yet the fact that he does bark is one of the most wonderful things in the jungle. It is so wonderful that in another book I shall tell you more about it. But now I shall tell you just this:

There are some animals which are so deadly that they could kill off many, many other animals. So, as the only way to save the other animals from being all killed, *God has made some special animals to fight those deadly animals*.

There is the *cobra,* which is a snake, and which has such a deadly poison that it could kill almost all other animals in the jungle by just [115] biting them. So, to save the other animals from being killed by the cobra, God made the *mongoose.* He is a plucky little creature, about the size of a cat. And he will fight and kill every cobra he sees! But really he is such a wonderful animal that I must keep him for another book, when you are old enough to know him better and to love him.

But sometimes the deadly animal is too strong to be killed himself. There is the tiger. He can kill and eat many kinds of animals. But who can kill *him*? No animal! At least, the elephant and the buffalo could kill the tiger if the tiger should let them *catch* him and trample on him. But the tiger does not let any animal catch him. Then how can the other animals be saved from the tiger?

God made two special animals to save the others from the tiger. The first is the buffalo, of which I have already told you, and which is the Knight of the Jungle. The second animal is the barking deer. How the barking deer saves the other animals from the tiger, I shall now tell you:

When the tiger is prowling about, all other kinds of deer and antelope just run away, and are glad enough if they escape being eaten. [116] But not the plucky little barking deer! He too runs away, but as soon as he gets a little ahead of the tiger, he stops under a bush and lets out that bark or yap—then runs on at once to another bush.

The tiger is furious, and jumps on the bush where he heard the bark—but the deer is not there now! The deer barks from that second bush—and runs to another one. In this way the barking deer leads the tiger on and on through the jungle from bush to bush.

And why does he bark like that? To tell the other animals in good time that the tiger is coming, and then to tell them exactly *where* the tiger is.

"Look out, here's a tiger!" That is the meaning of his first bark.

"Here he is! He is coming after me—this way!" That is what he means by the next bark.

"He is chasing me this way! You run the other way!" And that is what the barking deer keeps on saying, as he runs from bush to bush, so that all the other animals know exactly where the tiger is at each minute.

In this way the barking deer runs through the jungle, *warning all the other animals*, and so spoiling the tiger's dinner all the way.

[117]

CHAPTER X

Deer and Antelope: Their Special Gifts

You have learned by this time that *every animal has some special gift*, that is, he can do one thing better than most other animals. The deer and the antelope have their special gifts.

First, there is their gift of *hearing*. I have already told you that the wild buffaloes can hear a long way; but the deer and the antelope can hear still farther.

Let us suppose that a tiger is trying to creep up to a deer through the jungle, as quietly as he can. The tiger is still a long way off, and quite hidden by the bushes, so the deer cannot *see* him at all. But the deer can *hear* him coming, even if the tiger takes each step very lightly. Why? Because the deer's ears are so sharp that he can hear even a leaf rustling under the tiger's foot, a long way off. So the deer can run away in good time.

To make him hear still better, the deer can turn or bend his ears to the side from which the sound is coming. You have seen an ordinary [118] cow prick up her ears when she heard somebody coming; and many other animals—even a dog—can do the same.

But the deer can do that best. The shape of his ear is like that of a funnel, so as to *pour* the sound into his ear, as it were. Then even if there is only a single drop of sound, it gets right into his ear.

And by turning or bending his ear, the deer knows which way the sound is coming. You also can tell which way a sound is coming, if it is loud enough; but the deer can do that even when the sound is very faint. That is very useful to him, as he then knows exactly *which way* a sneaking tiger is coming, and can run the *other way*.

I must now tell you that the tiger himself, tries to come so quietly that the deer may not hear him at all; and to help him to do so, his feet are padded with muscles, just like cushions. So it is a kind of trial between the tiger and the deer as to which is the more clever. If the tiger can come so quietly that the deer cannot hear him, then the tiger is more clever than the deer. But if the deer can hear the tiger, even if the tiger comes most quietly, then the deer is more clever than the tiger. [119]

That kind of trial between two different animals as to which is the more clever, goes on in the jungle all the time: and *the more clever one wins every time*. If the tiger is more clever than the deer, the tiger eats the deer; but if the deer is more clever than the tiger, the deer escapes being eaten. And that is true of all other animals. In fact, one of the great wonders of the jungle is that the animal which is *the fittest wins the oftenest*; and so he goes on living, whatever may happen to the others. [1]

[1] *To the Teacher.*—Please give the class other examples of the "Survival of the Fittest" among other creatures—birds, insects, fish, etc.

Now I come to the second special gift of the deer and the antelope. If by any chance a deer cannot hear a sneaking tiger, he can still *smell* the tiger.

Most animals can smell their enemy a long way off, even if they do not hear him or see him; but *the deer and the antelope can smell the farthest*. Even if a sneaking tiger is so cunning that he stops in a thicket and stands quite still for a minute, so that he does not make any sound at all,—and so the deer cannot hear him,—even then the deer can smell him when he is still a long way off.

[120]

I must tell you now that the tiger himself can smell the deer. But he cannot do that very far off,—so the deer always smells him *first*!

Also, the tiger can hear the deer, if the deer happens to be moving. But the tiger cannot hear quite so far as the deer can. So the deer always hears him *first*!

But in one thing the tiger is better off than the deer: *the tiger can see farther than the deer*. In the night most animals can see a little, but the tiger can see a little better and farther than the others. And in the daytime, if a deer were feeding in a very big level field, and a tiger came to the field from the other side, the tiger would see the deer before the deer could see him. Then the tiger would come round to the nearest thicket, and try to creep up to the deer from thicket to thicket.

Each Animal has the Gift he Needs Most

So, you understand, the deer can *hear* farther and *smell* farther; but the tiger can *see* farther.

And that is so because it is a wonderful rule in the jungle that *each animal has the gift that he needs most*.

But can you think why the tiger *needs* to [121] see farther, and why the deer *needs* to hear farther and smell farther? I shall tell you.

The tiger is the catcher, and the deer is the one that is caught. So the tiger tries to get to the deer, and the deer tries to run *from* the tiger.

But to get to the deer, it would be no use to the tiger if he could only smell or hear the deer, for then he would only know that the deer was *somewhere* near, but could not find the exact spot; and to catch the deer the tiger must know exactly where the deer is. So the best way for him to know that is to *see* the deer.

But for the deer himself, all that he needs to know is that a tiger is somewhere near. So it is quite enough for him to know from which side the tiger is coming, by just smelling him or hearing him. Then the deer can run the other way at once. He does not want to see the tiger at all!

So, you understand, the tiger's best gift is to be able to see the deer; and the deer's best gift is to be able to smell and hear the tiger.

But then, you may ask, if the deer can always run away long before the tiger can get at him, does a tiger never catch a deer?

Yes, a tiger does catch a deer once in a while, [122] if the deer happens to make a mistake! And the deer can make only one mistake like that in his life, because after the first he gets eaten!

So, you may be sure, the deer tries very hard never to make even that one mistake.

And what is that one mistake? It is to run straight into the jaws of the tiger! It may just happen that when the deer hears the tiger coming, he does not listen quite carefully, and so he does not know which way the sound is coming. Then, in running away, the deer may happen to go just the wrong way—and fall into the tiger's jaws.

Or else it may happen that the deer is so frightened that he loses his head, as it were, and goes just any way—and by bad luck chooses the wrong way, and falls into the tiger's jaws.

But I must tell you that, although the tiger tries very hard to eat the deer, *the deer tries still harder not to be eaten!* Why? Because if the tiger does not catch the deer for to-day's dinner, he can still catch some other animal for tomorrow's breakfast, even if he goes hungry to-night. But if the deer once gets eaten, there is no to-morrow for, him at all! The tiger is only trying *to get a meal,* but the deer is [123] trying *to save his life.* That is why the deer nearly always gets away from the tiger—because he is trying harder than the tiger.

So the tiger does not get deer to eat much oftener than most children get roast turkey. The tiger lives mostly on pork, for the wild pigs of the jungle are such careless animals, as I have told you before. Now and again the tiger gets mutton also, for the wild sheep are silly creatures, like other kinds of sheep. In the same way the tiger sometimes catches a wild goat.

The tiger would really get deer to eat a little oftener than he actually does if it were not that the deer has two other gifts by which he can escape from the tiger at the last minute. Those two gifts are his *quickness in getting started*, and his *speed in running*. So, even if the deer makes a mistake and runs toward the tiger, he can still escape from the tiger if he finds out his mistake in time.

For, as you saw at the midnight pool, the deer may be drinking quietly, when he hears or smells a tiger. Then the deer can leap at once and get away, before the tiger can leap. Or it may happen that the deer is trying to escape from a tiger and has run to within [124] twenty yards of the tiger, when he finds out his mistake. Then the deer can turn *at once* and leap sideways to get out of the tiger's reach. The deer is so quick that he can turn aside without stopping, and keep on running.

Then after that, once he has turned away from the tiger, the tiger can never catch him. For the deer can run ever so much faster than the tiger.

In fact, the deer or the antelope is the fastest animal in the world, except one other. About that other animal I shall tell you some wonderful things in the next book. But among all animals I have told you about in this book the deer is the fastest.

"But how do people know that the deer can run faster than other animals?" you may ask. "Has anyone had a race between different animals?"

Yes, some people did that in England a few years ago. They took the fastest racehorse in the country, and ran a race between him and the fastest greyhound; and the greyhound beat the horse in the race. Then they took that greyhound, and ran a race between him and an English deer; and the deer beat the greyhound in the race. So, you see, the deer [125] was faster than the greyhound, and the greyhound was faster than the horse! So the deer was the fastest of the three.

And the deer that lives in the jungle is even faster than the English deer. Why? Because the English deer lives in peaceful glades and forests, and has no other animal trying to catch and eat him; so he does not try to be as fast as he could be. But the deer that lives in the jungle has to try very hard all the time to be as fast as he can be, or else he would be eaten by the tiger! And, as you must know, *we can do the best in anything when we try the hardest*.

So, all kinds of wild deer in the jungle have been trying their hardest to run as fast as they can. And as their fathers and grandfathers have been trying to do that, the wild deer to-day have become the fastest runners among all the animals I have told you about.

[126]

CHAPTER XI

The Camel

The *camel* has very little to do with the kind of jungle I have been telling you about; but he has much to do with the *desert*. A desert is another kind of wild place. As I told you before, jungle means any wild place; but usually, of course, there are lots of trees and bushes and thickets in it. But we call the wild place a desert when trees and bushes and thickets will not grow there, because the ground is all covered with *sand*. In the desert there is nothing but sand all over the ground, and not a single tree or a tiny blade of grass anywhere, as far as you can see.

And that is the place where camels can do some very wonderful things, as I shall now tell you. The camels do not actually live in the desert all the time, but in countries quite near there.

First I must tell you that there is only one country to-day, called Central Asia, where camels are still found wild. In all other places they [127] are not wild any more, for in those countries people have lived for many thousand years; so the people caught all the camels once upon a time, and tamed them.

Since that time the camels have been used by people in those countries for their work, just as we use horses here; and rich people in those countries count their wealth by the number of camels they have. Just as we say here that a rich man has a million dollars, or two millions, or three millions, so in those countries a man is thought to be rich who has one thousand camels, or two thousand, or three thousand.

It was just the same in those countries in olden times. You have read in your Bible history that Job was once a rich man, as he owned thousands of camels.

You will see from the pictures facing page 128 that there are two kinds of camels; one kind has a huge hump on the middle of his back; and the other kind has two humps, with a gap between. The *One-Hump camel* is called an *Arabian camel*, or a *dromedary*. Once upon a time he lived in the country called Arabia; that is the country from where you get your lovely old stories of Ali Baba and Aladdin. But now the One-Hump camel also lives in other countries near there. These are all very hot countries, with many miles of desert here and there.

The *Two-Humps camel* is called a *Bactrian camel*, as he lives in a country which was once called Bactria. That country also has many deserts, like Arabia; but as it is far to the north of Arabia, it is very cold in winter, and the snow then lies very thick on the ground. So try and remember this:

The One-Hump camel lives in a country where there are many miles of desert, and where it is very hot almost the whole year. So the One-Hump camel has to guard himself only from the *hot burning sand*.

The Two-Humps camel lives in a country where there are also many miles of desert, but where it is very hot in the summer and very cold in the winter. So the Two-Humps camel has to guard himself from the *hot burning sand in the summer*, and from the *cold and snow in the winter*.

The Two-Humps camel has in winter a coat of long, shaggy hair on his back to guard him from the cold; and in summer the shaggy hair comes off his back, just as if he were to cast off his thick coat. But the One-Hump camel has only short hair, as the country is too hot all the time to need a thick coat.

Bactrian Camel — with Two Humps

Arabian Camel — with One Hump

[131]

Now I must tell you how camels are used. First, they carry goods for trade. In those countries there are hardly any railroads, so the merchants carry their packages on camels. Of course they could not put a package right on a camel's hump, as it would fall off; so they always join two packages together with a band or belt, and sling the band across the camel's back, so that there is a package on each side of the camel.

When a One-Hump camel is used, the band or belt has two parts, like a loop; and the loop rests over the hump, so that the band cannot slip backward or forward. When a Two-Humps camel is used, the band of course rests in the gap between the two humps, so that it cannot slip at all; and then the two packages can be made very big. That is why people like the Two-Humps camel better for carrying goods, and like the One-Hump camel better for riding. But in some places the One-Hump camel is used both for riding and for carrying goods.

In this way merchants carry their goods for many hundred miles across desert and country. Then sometimes they come to the sea and send the goods in ships to different countries. That is how you get many of the figs, dates, [132] and grapes you eat; so the next time you eat them, think of the patient camel that brought them for you across the desert. That is why the camel is called the *Ship of the Desert*.

The beautiful carpets and rugs and shawls which you see in rich homes have also been brought by the patient camels; and some of the lovely vases and ornaments that rich people have were also carried by camels. And not only across the desert, but even over ordinary land camels carry these goods. The camel is such a large animal that he can carry packages as heavy as four men.

Of course when he carries such a heavy load, he cannot go any faster than a man's walk; but the camel can keep on walking all day, with just a short rest once in a while. Those used for riding cannot run as fast as a horse, but they can keep on running at a steady trot much longer than a horse, and then after a short rest can start running again. So by the end of the day a camel can run twice as far as a horse, and sometimes still farther.

The Camel's Wonderful Gifts

Now I am going to tell you of the most wonderful things a camel can do. [133]

First, I must tell you that no other animal could cross a desert at all. To begin with, if such an animal as a horse tried to walk on the sand, his hoofs would sink into the sand up to the ankles, and it would be hard work for him to go even a mile. But a camel's foot is different. It has a *soft pad of muscles* under it, just like a cushion; and when the camel walks or runs on the sand, the pad spreads out under his foot, and that gives him a firm hold on the sand in walking or running. So remember that the camel has padded feet.

I must tell you here that the feet of all animals are formed in the way they can best use, in the country in which they live. Those animals that have to walk on *hard ground* have *hoofs*, and those that have to walk on *soft ground* have *padded feet*. The elephant is the only animal that has to walk on hard ground, at least very often, and yet has padded feet. Can you tell why? Because of his huge weight! He is so heavy that if the bones under his feet were not covered with a thick pad, he would jar the bones every time he put his foot down, even if the ground were not very hard.

In the same way the camel's padded foot is very useful to him even when he is not in the [134] desert, but on hard ground; for he too is rather heavy, though of course not so heavy as an elephant.

Sand Storm in the Desert

There are other reasons why no other animal could cross a desert as easily as a camel. In the desert there are sometimes fierce storms; and as it is all sand there, the strong wind blows the sand about in every direction. As there is no place there where one could get away from the sand, any other animal would soon have a lot of sand blown into his nostrils; then he would be choked. But a camel's nostrils are made differently, so that whenever he likes he can *close his nostrils*, as easily as you can close your mouths, and that keeps away the sand. The camel is clever enough [135] to lie down on the ground when a storm is blowing, and to lay his neck and chin along the ground; then his nose is quite close to the ground, where the storm is not quite so fierce as in the air.

Of course, when he wants to breathe, he opens his nostrils a tiny bit to take in a little air; then he closes the nostrils again, and holds his breath for a little while. He has to keep on doing that as long as the storm lasts.

But what does his master do, who has been riding on his back? He cannot close his nostrils; so the only thing he can do is to get off the camel and huddle against the camel's body on the side far from the wind; then he brings his face quite close to the ground and holds his nose with his hand. When he wants to breathe, he opens his fingers just enough to make a slit and let the *air* in, but not enough to let the *sand* in.

There is another reason why no other animal could cross a desert: his eyes would be blinded by the fierce glare of the sun. But a camel has very *thick hair on his eyebrows*, which hang over the eyes, and keep off the fierce rays of the sun. His eyelashes also are very thick, and help to keep off the sun in the same way. [136]

But there is a still more wonderful reason why no other animal, except a camel, could cross a desert. In a desert water is very scarce, and a traveler crossing a desert on a camel may not find any water for a whole week. Of course, he carries on his saddle half a dozen bottles of water to drink; but after drinking some of the water each day, he has not much to spare for the poor camel. Then what is the poor camel to do?

Of course, you may say that his master should carry more water for the camel to drink. But the load of goods which the camel has to carry is already so heavy that there is not much room for any more water. Then what *can* the poor camel do?

Why, he *carries his own drinking water*, not in the load on his back, but *inside his stomach*! Is not that a wonderful thing? His stomach is made differently from that of any other animal. The stomach of any other animal, Or even a man's stomach, is so made that the water drunk at any time is all used up in the next few hours; that is why any other animal, or even a man, has to have another drink after those few hours.

But a camel's stomach is so made that it [137] has one big place for food and drink like the stomach of any other animal, but it also has many smaller places arranged all around the stomach; these smaller places are just like bottles, and are called *cells*.

So when a camel takes a good long drink, the big place in the middle of the stomach takes in the water first; then as he drinks more and more, the bottles or cells all around begin to get filled also. And the wonderful thing is that as soon as each cell is full, its mouth closes up by itself! In that way, if the camel drinks long enough, all the cells get full, one by one, and then have their mouths closed up.

When a camel is about to start on a long journey through the desert, he takes a very long drink, till he *feels* that he cannot drink any more; then he *knows* that all the bottles or cells inside are quite full, as well as the big place in the middle of his stomach. Now he is ready to cross the desert.

After many hours all the water in the big place in the middle of the stomach gets used up. Then what happens? Why, one of the bottles inside opens its mouth by itself, and pours the water into the stomach! And after many hours more, when *that* water has also [138] been used up, the *second* bottle opens its mouth and pours the water into the stomach. In this way all the bottles or cells inside the camel one by one pour their water into the stomach from day to day, whenever the camel feels thirsty. Is not that most wonderful?

And there is yet another very wonderful thing about the camel. His hump! It is just as wonderful whether it is one hump or two humps. I shall tell you.

The camel's hump is his *store of food*! Yes, just as he carries his own drinking water inside his stomach, so he also carries his own store of food in his hump.

This is how he does it:

When the camel is quite well and strong, if he eats any food which is a little more than he actually needs for his hunger, that food after a while goes to his hump and helps to make it bigger. In this way the hump becomes a store of all the extra food that he has eaten. Then, on going on a long journey through the desert, if the camel has nothing to eat and begins to feel hungry and weak, a little of the hump is used up to give him strength, just as if he were to eat a meal. In this way he can go for many days without [139] food, but of course his hump will get smaller and smaller.

Crossing the Desert with Camels

But his master does not actually take him through the desert without giving him *any* food or drink; in fact he always gives the camel some of the figs and dates which he takes with him for his own meals, and also some of the drinking water which he carries on his saddle. But if it did happen that his master had no food or drink to spare, the camel could still live for several days, using his hump for food, and the water in the cells of his stomach for drink.

The camel can do yet another wonderful thing. He can tell a long way off when he is coming to a place where there is water. In [140] the desert, after going over sand and sand for many days, a traveler sometimes finds a beautiful place called an *oasis*. It is just like a lovely little garden right in the middle of the desert, with a spring of water, and several fig trees, date trees, and other palm trees growing all around the pool.

When a traveler is crossing the desert and sees nothing but sand for several days, it sometimes happens that his camel suddenly stops, stands quite still for a minute, raises his head, and sniffs the air. Then he turns a little to the right or to the left, and begins to run straight that way. His master may look ahead very hard, but he will see nothing but sand and sand, as before.

But the camel, by just sniffing the air, has found out that there is an oasis within reach, though it is still too far away for him to see it. Then he runs on most gladly, and comes to the oasis in an hour, so that he and his master can rest there for some time, and drink from the pool, and eat the figs and dates growing on the trees.

Of course, the camel can also eat the leaves of the trees; in fact, when he is not in the desert, but just in the ordinary country, he [141] usually eats from the shrubs and bushes, and gets figs and dates only as a dainty, just as you sometimes have ice cream. The camel with two humps will gladly eat many more things than the camel with one hump. In fact, when he is hungry, he will eat not only any kind of vegetable, but also meat. He has even been known to chew up and eat bones, blankets, and leather! And he is perhaps the *only animal that will drink salt water*; for the country in which the Two-Humps camel lives has several lakes, the water of which is bitter and salty.

So you see how useful an animal the camel is, whether he has one hump or two humps. He is so useful that people have been saying for a long time that camels should be brought into America, where there are several deserts in the western states. In fact, a strange thing has already happened. The United States Government did bring a lot of camels for use in the western states several years ago, about the time when your grandfather was a boy.

But the people who can best manage such large animals as elephants and camels are the people who are born in the same countries as those animals and who understand their habits. And unluckily, when the camels were brought [142] into America, nobody thought of bringing men also from those countries to manage the camels. So nobody seemed to know how to use these animals, and after a time they were turned loose in Arizona. The camels went into the deserts and forests there, and became quite wild, and today there are some of them in Arizona.

Now, do you not think it would be a good idea to get a few men from those countries and learn from them how to manage camels? Then the camels of Arizona also could be used in crossing the deserts there, where there are no railroads.

Besides being the only animal that can cross the desert, the camel is different from any other four-legged animal even in the way he walks.

You have seen how a horse walks? When his left foreleg is lifted off the ground, his right hind leg is also lifted off the ground; then in the next step, when his right foreleg is lifted off the ground, his left hind leg is also lifted off the ground. That means that the two legs which move at the same time are those placed at the *opposite corners* of his body. But when a camel walks, he lifts the [143] two legs on the *same side* of his body at the same time. And when he takes the next step, he lifts the two legs on the other side of his body.

Now, my dear, I have told you many things about the camel which are different from anything in any other animal. So, before I close this chapter, I want you to remember these things about the camel:

1. His *foot* is *padded* in such a way that he can walk or run on sand.

2. He can *close* his *nostrils* to keep out the sand in a storm.

3. His *thick* and bushy *eyebrows* and *thick eyelashes* keep the glare of the sun from his eyes.

4. His stomach has many *cells* like bottles, in which he can *store up water*.

5. He can *store up food* in his *hump*.

6. He walks by moving both legs on the *same side* of his body at the same time.

In another chapter I shall tell you about an animal that can also do one of these things: he can *store up food* in his body, though in a different way. That animal is the *bear*. He sleeps through the whole winter, and has to have a store of food somewhere in his body to last all that time.

[144]

CHAPTER XII

The Camel and the Thief

Now I shall tell you a story about a camel and a thief. It is a true story, and happened many, many years ago. The story shows what we can learn by watching the animals.

Once upon a time, a traveler was going on foot across the country. In his belt he had a purse full of money. One day, as the sun began to get hot, he lay down on the grass under a tree near the roadway, and fell asleep.

After a few hours he woke up, and what was his surprise to find that the purse was gone! While he was asleep, somebody had quietly stolen his purse and gone away.

The traveler ran to the nearest village, and there told the police about it. Now, among the police there was a very clever man, and the police brought him with them to the place where the money had been stolen. The clever man looked all around the place very carefully to see if he could find any marks on the ground. On the grass near the tree he found no marks; in fact, if a person walks on the [145] grass just once or twice it does not leave any mark. But on the roadway near by he found footprints.

"They are a camel's footprints," he said, looking at the marks carefully. "And the marks of all the four feet are not quite the same. Three of them are quite deep and clear; but the fourth one is very faint."

He followed the camel's footprints along the road for a long time. But now and again he stopped and looked at the shrubs and bushes which grew here and there, on both sides of the road.

"Hello, that is strange!" he suddenly said. "The camel has eaten from the bushes and shrubs here and there on the left side of the road, but he hasn't eaten at all from those on the right side of the road."

He went on for some time longer, then suddenly stopped to look at the road where the camel had walked.

"Hello, this is also strange!" he said. "Here are a lot of *bees* buzzing near the ground on the *right* side of the road. And here are a lot of *ants* scrambling over the ground on the *left* side of the road."

"Never mind about the camel, and the bees [146] and the ants," the policemen said impatiently. "We want to know about the thief who stole the money. You have not found any other footprints except the camel's?"

"That is quite true," the clever man said. "But as the *camel* could not steal the money, there must be a *man* riding on the camel. He must be the thief."

"But why didn't the thief leave any footprints?" the policemen asked.

"Because he must have ridden his camel from the roadway right to the edge of the grass," the clever man answered. "Then he must have jumped down upon the grass, where he knew he would not leave any footprint. He must have walked very quietly on the grass up to the tree where the traveler was sleeping, and stolen the money. Then he must have walked back quietly to the camel and ridden off."

"But what sort of a man is the thief?" the police asked. "How can we find him, if you do not tell us what he is like?"

"I cannot tell you a thing about the thief, or what he looks like, as he hasn't even left a footprint," the clever man answered. "But I can tell you *all about the camel*. The camel is [147] *blind* in his *right eye*, and *lame* in his *left hind foot*. And on his back he is carrying two packages, one on each side; the package on the *right* side has *honey* in it, and the package on the *left* side has *corn* in it. So you must search for a man who is riding a camel loaded like that. He is the thief."

So the police searched for a man who was riding a camel which was blind in his right eye, lame in his left hind foot, and carrying honey in a package on his right side, and corn in a package on his left side. After following the camel's footprints on the ground for a long time, the police at last came to a village.

They searched through the village, and found many men riding camels. But there was only one man riding a camel blind in his right eye, lame in left hind foot, and carrying honey on the right side, and corn on the left side. So the police knew that he was the thief, and took him before the judge. Then the thief said that it was quite true that he stole the money.

Afterwards the judge turned to the clever man and asked him how he knew all that about the camel.

"You didn't *see* the camel at all, but only [148] his footprints," the judge said. "Then how did you know that the camel was blind in his right eye, lame in his left hind foot, and carrying honey on the right side, and corn on the left side?"

"It was quite simple," the clever man answered very modestly. "First, about the camel being blind in his right eye. He had nibbled at the shrubs and bushes growing on the left side of the road, for at each bite I found the leaves cut off clean by his teeth. On the right side of the road there were also plenty of good shrubs and bushes, but the camel had not taken a single bite at any of them. That showed that he did not even *see* those shrubs and bushes on his right side. And that of course meant that his right eye was blind."

"That is very clever of you," the judge said. "But how did you know that the camel was lame in his left hind foot?"

"That was just as simple," the clever man again answered very modestly. "As the camel walked along, the marks of his two front feet and right hind foot were quite deep and clear on the ground. But the mark of his left hind foot was very faint. That showed that the [149] camel was limping, and the left hind foot only just touched the ground. So I knew that he was lame in that foot."

"That is also very clever of you," the judge said. "But how did you know that the camel was carrying honey on his right side, and corn on his left?"

"That was the simplest of all," the clever man answered most modestly. "As the camel was limping, nearly every step he took jerked the load on his back. So a few drops of the honey fell to the ground from the package on his right side, and a few grains of the corn fell to the ground from the package on his left side."

"But you could not see very well here and there on the ground just a few drops of honey or just a few grains of corn?" the judge said.

"*I* could not," the clever man answered, "but the *bees* and the *ants* could! On the right side of the road I found a swarm of bees here and there; so I knew that they were trying to pick up the honey. And on the left side of the road I saw a whole lot of ants here and there; so I knew that they were trying to pick up and carry away the grains of corn." [150]

Now was it not really clever of that man to find all that out about the camel, without ever seeing the camel before? But, as you understand, he knew all about the *habits* of different animals; and so he knew what camels and bees and ants always do.

[151]

CHAPTER XIII

Bears

Bears are such funny animals, at least some kinds of bears, that you may like to know all about them.

Bears are found in many countries, and in some countries there are several kinds of bears.

But you must remember this: *hardly any bear lives in the tropics*; that means countries where the sun is almost overhead all the months of the year, so that it is very hot all the time.

But why does not the bear usually live there? Can you guess?

Because the bear is a very *hairy* animal; and his hair is just like a thick coat, so that he cannot live where it is very hot all the time.

Of course, once upon a time the bear lived only in places where it was very cold, and so he grew thick hair to keep out the cold; but now that he *has* a thick coat of hair, he cannot go down to hot countries to live. He does not mind living in a cold country; and the colder the country is, the thicker is his coat of hair. [152]

What does the bear eat? Most kinds of bears eat berries, fruits, soft roots of trees, and fish when they can catch it. One or two kinds of bears eat other things also, which I shall tell you about very soon.

The bears that live in cold countries, where there is heavy snow in winter, cannot get anything at all to eat in winter. Why? Because there are no fruits and berries in winter, and the roots of trees are frozen hard and covered up by the snow.

Then if those bears cannot get anything to eat in winter, what do they do? They *sleep*!

You know that when you are asleep you do not feel hungry; but as soon as you wake up you feel hungry again. It is just the same with the bear; he does not feel very hungry while he sleeps. And he *sleeps right through the winter months*!

Still, while he is asleep all that time, does he not feel a little bit hungry? He does. So he uses up the store of food inside his body! I have told you that the camel carries a store of food in his hump. The bear has no hump, of course, but he has a *thick chunk of fat all around his body* just under his skin; and that chunk of fat is his *store of food*. [153]

So, when the bear sleeps snugly in his den in the winter months, the chunk of fat is slowly used up inside his body, and keeps him from being very hungry.

Of course, he eats such a lot just before the winter, that the chunk of fat is very thick when he goes to bed. But the chunk is all used up when he wakes up at the end of the winter, and then he is very hungry again!

But there is a kind of bear that lives in a place where there is snow and ice almost all the time. What can *he* do?

He cannot sleep always! So he has to get something to eat now and then, and I shall tell you how he does that.

The Polar Bear

This kind of bear is called the *polar bear*. (See the picture on page 155.) He lives in a place far up North, where it is always very cold. The land is nearly covered with snow, and the water at the top of the sea is frozen. There are no berries or fruits there for the polar bear to eat; so he has to live on fish, and seal, which is a water animal. The way the bear catches the fish or the seal is this:

He makes a hole in the ice with his paws, so [154] that he can reach the water below. Then he sits down very quietly by the edge of the hole, and waits for a fish or a seal to swim past the hole. Then the bear pounces on it very quickly with his paw or his jaws, and catches it.

If the ice is too thick for the bear to make a hole through it, he has to try another way. He comes right down to the part of the sea where some of the ice has broken off. There he chooses a place at the edge of the ice, close to the water; and he waits there for a fish or a seal to swim past. Then he pounces on it and catches it.

Now I shall tell you a few special things that the polar bear has.

His coat of hair is much *thicker* than the coat of any other bear. Why? Because he lives in a colder place than any other bear; so he *needs* a thicker coat. Also, he sometimes has to swim through the icy water to get to some floating field of ice, so that he can catch fish from it. Then, although his hair gets wet, he has a thick lining of fat inside his coat to keep him warm.

The next special thing about the polar bear is that his hair is *all white*—like the color of everything around him, which, as I have told [155] you, is just snow and ice. So when the polar bear sits down very quietly on the snow and ice, nobody can see him even from a short distance, because he is the same color as the snow and ice. And that is why the fish or the seal does not see him, and so gets caught.

Polar Bear

That is one of the wonderful things about many wild animals— they are of the *same color as the place where they live*. You know that the color of a lion is yellow, like the color of sand; and the lion lives in countries where there are lots of sandy places. You know, too, that the [156] color of a tiger is yellow, but with black stripes upon the yellow, so that if you looked at him from a distance, you might think he was made up of yellow and black stripes. And the tiger lives in the tall grass, which also looks like yellow and black stripes.

But now I shall tell you more about the polar bear. He has three other special things: the *soles* of his feet are *hairy*; he has a *small head*; and he has a *long neck*.

First, about the soles of his feet. The soles of the feet of other bears are smooth, just like the feet of all other animals that have to walk on ordinary ground. But the soles of the polar bear are covered with long hair, just as is his body. Why? Because he has to walk on ice, which is very slippery, and he needs to have the soles of his feet covered with hair, or else he would slip on the ice, just as you must wear rubbers over your shoes when you have to walk on icy ground.

Now, my dear, just stop for a minute, and think. Among all the wonderful things that I have told you so far, you have always noticed that an animal always has *just the very thing he needs! We* have to *make* rubbers, and warm coats, and gloves, and socks, and a dozen things [157] that we need. But to every animal God has given everything that he needs, right on his body.

But now let us go on with the polar bear. He has a smaller head than any other kind of bear. Why? To make it easier for him to put his head through the hole in the ice, when he is catching fish. Other kinds of bears do not have to put their heads into a hole to get anything to eat; so they do not need to have a small head.

The polar bear has also a longer neck than any other kind of bear. Why? To give him a longer reach in catching the fish with his jaws—without tumbling into the water himself. Other bears, who live on dry land, do not need to reach out like that, and so they have shorter necks.

I shall now tell you about these other kinds of bears.

American Bears

First you shall hear about the bears that live in America. The biggest kind is called the *grizzly bear*. In fact, he is the largest bear in the world. Some grizzly bears are ten feet tall when they stand up on their hind legs!

The color of a grizzly bear is yellow, but with [158] many shades; sometimes between brown and yellow, and sometimes between red and yellow. Teddy bears, with which you have played, are sometimes made of that color. Teddy bears of course are very nice, as they are toys; but I am sorry to say that the real grizzly bear is not nice; he is very fierce. In fact, he is the only kind of bear that is so fierce. Even grown-up men do not want to go near him.

The grizzly is found in many parts of North America, near the Rocky Mountains, from the United States right up to Alaska. He lives on berries and all kinds of fruits, and also on the soft roots of trees. But the grizzly bear eats meat also, if he can manage to catch deer or cattle. That is why cowboys in Colorado and Wyoming do not like the grizzly bear—he tries to kill and eat their cows.

Besides, he kills a lot of fish. In the Columbia River in Oregon there are lots of trout and shad, which people like to have for their dinner. But the grizzly bear gets to the river first, and eats a great many of the trout and the shad. How does he catch the fish? Why, he just lies down along the bank, and waits for the fish to rise to the top of the water. The trout and the shad like to rise to the top of the water now [159] and again, and swim there. So the grizzly just waits for a fish to rise—and then he pounces on it and catches it with his paw. He is so very quick that he hardly ever misses. All kinds of bears are very clever in catching fish.

Other Bears

Another kind of bear is called the *brown bear*. He lives in Europe, Asia, and also in some parts of America, especially in Alaska. There he is rather big, though not quite so big as the grizzly bear. He too lives on berries, fruits, and roots, and he also catches fish. For in the rivers of Alaska there are lots of salmon.

But the brown bear is not at all fierce, like the grizzly bear. He is peace-loving, and sometimes quite friendly.

The nicest kind of bear is called the *black bear*. He is found in all parts of the United States, and in many other countries. He is sometimes rather small, and is quite full of fun. Almost all the good stories you may hear about bears are about the black bear.

Many people mistake the black bear for the brown bear; so when you sometimes hear people talking about a "brown" bear, you may know that they really mean a "black" bear. [160]

Like other bears, the black bear lives on berries, fruits, and roots, and also on nuts, if he can find any. But what he likes best is honey! It is quite amusing to see the bear hold a honeycomb in one paw, scoop out the honey with the other, and put it into his mouth. It looks just like a boy holding a pot of jam in one hand, and sticking his fingers into the jam and putting it on his tongue!

"But do not the bees get angry, and try to sting him?" you may ask.

Of course they do. The bees swarm around the bear and try to sting him all over. But they cannot! He is *too hairy*! They cannot get through the hair to sting him on the skin. So he goes on licking the honey and smacking his lips!

The black bear is always a funny animal. Perhaps you have seen him in the zoo. He will squat on the ground like a man, and if he sees a crowd of people before him, he will swing his arms just as a man does when he talks to a friend. Perhaps the bear has seen some men do that, and has learned to do the same!

And sometimes he will sit on the ground, hold his tail in his mouth, and fumble head over heels, or roll over and over, or spin round and [161] round — just for fun! In fact, the black bear is among the *few grown-up animals that love to play*. Many *young* animals of course, such as kittens, puppy dogs, calves, and many others, love to play. But most grown-up animals do not seem to care for play, except the black bear.

When he is caught and tamed, he is still very playful at times. He will do all sorts of funny tricks, all by himself; and if he sees anyone watching him, he will try to show how clever he is—just like a child playing "smarty!"

Once in a lumber camp in the West the men caught and tamed a black bear. After a time the bear was allowed to walk about the camp, quite free, as he did not think of running away. One day the men had stopped work to have their dinner. The bear walked by, and the men petted him and said nice things to him. Then what do you think happened! The bear felt so proud of being praised that he went to a sloping log, and walked along it right to the top. You must know that in a lumber camp there are lots and lots of huge logs, or trees which the men have cut down. And one of these logs happened to rest on a slope, that is, with one end higher than the other.

When the bear reached the top of the log, he [162] sat across it. Then he held on to the log with his front paws, bent his body, and slid down the log—just as a boy slides down the banister! Of course the men laughed, and cheered him. Then just guess what that bear did!

He walked up to the top of the log again, and sat across it, as before. But now he held on to the log with his knees, not his paws, and sat straight up without bending, and slid down the log in that way—just as a boy might hold on to the banister with his knees, not using his hands at all, and slide down the banister in that way, just to show how smart he could be!

You may be sure those men cheered the bear, and gave him lots to eat.

There is another kind of black bear that is also funny, though in another way. He is called the *Himalayan black bear*, because he lives in India near some huge mountains called the Himalayas. In many ways he is very much like the black bear of America, but he has a *white chin* and *long side-whiskers on his jaws*. Some people think that of all kinds of bears he is the most handsome.

Although he often goes up very high on the side of the mountains, he sometimes comes down to the country below, where there are [163] many villages. But the bear is quite friendly, and never hurts the people in the villages, although he is strong enough to kill a man. So the people are also very kind and friendly to him, and never try to hurt him. When you grow up you may read that there are some people in India who are always kind to *all* animals, tame or wild. [2]

[2] *To the Teacher.* — Please explain to the class that the sect called Jains do not hurt the smallest creature, and will suffer the sting of a wasp rather than kill it.

Himalayan Black Bear

I am telling you this because you will see very soon what we gain by being friendly even to a [164] wild animal. The Himalayan black bear, like the other black bear, is also very fond of honey, and of everything sweet. In the country where he lives there grows a berry called mawa, which is very sweet — even sweeter than the strawberry; and the people of the villages make jam from it.

These berries grow quite wild, on bushes here and there in the fields, and even in the jungles near by. When the berries are ripe, the people send out their children to gather them from the bushes in the fields; and the children carry baskets so as to bring back as many berries as they can.

But when the berries are ripe, the bears also want to eat them! So it sometimes happens that half a dozen children are picking the berries from a thick bush, when suddenly a bear comes round the bush and starts gobbling up the berries as fast as he can!

Do the children get frightened and run away? Not a bit! They want *their* share of the berries, too!

By this time the bush may be getting empty, and the children have not quite filled their baskets. The bear keeps on gobbling up the berries, and even pushing past the children to [165] get at a bunch. What then? Why, the children raise their hands, and *just spank the wild bear*!

"Go away, you have had enough!" they say. "Can't you go to another bush? There must be others right in the jungle, where *we* can't go!"

And, can you imagine it, a wild bear there has never hurt a child! When the children spank him and push him away, telling him that he has had enough from that bush, he *does* go away to some other bush. Of course, the spanking does not really hurt him.

I have told this to you, my dear, just to show you that there is never any real reason for quarreling and fighting, among children, or even among men. *If children and wild bears can get along together*, why cannot children and children, or men and men, or nations and nations? Surely there are enough berries and other good things for all, if we only look around!

Remember this always, even when you grow up, if you want to be good men and women, and good citizens of your country.

[166]

CHAPTER XIV

Bears: The Tricky Trap

Now I am going to tell you something funny about the bear. You have seen lots of wild animals in the zoo, and you may sometimes have wondered how these animals were caught. In another book I shall tell you all about the different ways of catching different kinds of wild animals; but now I shall only tell you how a wild bear is caught. Of course, there are two or three ways of catching him alive, but I shall describe to you now just one way.

You must know by this time that everybody in the world—whether man, woman, or child—has *some fault*. Some have a bad temper, others are rude, and still others are obstinate; and many, especially children, are too greedy! And so it is among animals: they all have one fault or another.

So the people who want to catch a wild animal find out first what fault that kind of animal has—whether he is greedy, or obstinate, or bad tempered. And they *catch the animal because of that very fault*! [167]

A bear is very obstinate; in fact the bear, the pig, the donkey, and the mule are among the most obstinate of animals. So, because the bear is very obstinate, he will never give up when he meets anything that blocks his way; and if he has made up his mind to do anything, he will never give up, even if he finds he *cannot* do it and that it is very foolish to try to do it.

So the people remember the bear's obstinacy, and catch him in this way:

They find a large tree which has a bough fifteen or twenty feet from the ground; then they tie a pot of honey on the bough, quite two or three yards away from the fork where the bough joins the trunk. So, if a bear wants to get at the honey, he will have to climb up the trunk, and then walk along the bough to the place where the pot is tied.

But the people also take a heavy stone, tie a stout rope around it, and hang up the stone by the rope from another bough higher up. They place the stone in such a way that it swings right in front of the honey and a little above it. Then the people hide in thickets near by.

Presently a bear smells the honey from a distance, and comes to find it. On reaching the place he sees the pot of honey on the tree. [168] As the bear is a good climber, he soon scrambles up the trunk of the tree and walks along the bough toward the honey.

But just as he is coming to it, he sees something right before his path. It is the block of stone! And he cannot get at the honey without pushing the stone aside. So, what does he do? Why, quite naturally he pushes the stone aside with his paw. But, as I have told you, the stone is hung up by a rope; and so it *swings* any way you may push it.

Then what happens? Why, as soon as the bear pushes the stone aside with his paw, the stone *swings back* and hits him on the paw. The bear gives a growl, and again pushes the stone aside, and this time harder than before.

Then what happens? The stone swings back and hits the bear harder than before! In fact, the stone will always swing back just as hard as it is pushed.

But the bear does not know that! So with another growl he pushes the stone again—and now much harder than before. Then of course the stone comes back much harder, and whacks him again.

This makes the bear really angry. He hits at the stone, and sends it flying through the air [169] in a big curve. But when the stone has gone up and up in that curve, it begins to come down, down, the same way—and gives the bear a thumping whack on the jaw.

A Bear Fighting a Block of Stone

[171]

Now, if the bear were not such an obstinate animal, he would go away after that third blow, and try to forget the honey. But the bear will never, never, give in! Instead, he gets quite mad with rage. He thinks some enemy is hiding behind the stone!

"Who is hitting me?" he growls. "Come out of that, and fight fair!"

With that he hits a frantic blow at the stone; for the bear is a good boxer. He sends the stone swinging through the air again, and farther than before. Again the stone swings back and gives the bear a hard whack.

In this way the fight goes on. Of course the stone cannot get hurt; so it is the bear that gets hurt, every time. And as he will never give in, he goes on fighting with the stone, and gets hurt more and more, till at last he is knocked right off the tree, and falls stunned to the ground.

Then the clever people rush in from their hiding place, throw a net over the bear, and carry him away. And that is how the zoo gets some of its bears.

[172]

CHAPTER XV

Bright Birds

Now I shall tell you something about birds; not ordinary birds, but a special sort.

Of course, birds are not exactly animals of the kind that I have been telling you about, as they have only two legs, instead of four. But they have two wings, which are more useful to them than two more legs.

If they had four legs they could run fast; but with the two wings they can fly, which is ever so much faster and better than running. And they still have two legs with which to stand on the ground, when they have to come down to rest or to feed.

The birds that I am going to tell you about live wild in the jungle, and are free to build their nests where they like.

Among the birds we like best, some can *sing,* and some have *bright feathers.* Those that sing may live near your own homes in the country—the lark, the thrush, the nightingale, and some [173] others. But the birds that have bright feathers live generally in other countries.

Most birds that have lovely voices do not have bright feathers; and *most birds that have lovely feathers cannot sing.*

So among animals everything is very fair and just. With us it sometimes *seems* different. Some children appear to have all the good luck, and others all the bad luck. Some children can sing well, and are also very pretty; others cannot sing at all, and are also plain to look at. But really things are not quite so unfair; for a child who is plain, and cannot sing, may still have *some other gift.*

Among birds, those that can sing you may have seen often enough near your own homes in the country; so I shall now tell you about the birds that have bright feathers.

Most birds with bright feathers live in hot countries, where it is *sunny* almost the whole year. In fact, it is the bright light of the sun in those countries that gives the colors to the feathers of the birds, which are as lovely as the colors of the rainbow.

Among the bright birds that live quite wild in the jungles of hot countries, the most beautiful are the *flamingo*, the *parrot*, the *cockatoo*, the [174] *peacock*, the *golden pheasant*, the *egret*, and a few others.

The Flamingo

I shall tell you first about the flamingos, as they *live together in flocks*. They were once found in America, and only a few years ago there were many flocks of them in Florida, but now there are very few left in this country. They are now found in Africa and in the countries of southern Asia; a few are found also in Europe.

This is the way the flamingos live. They choose a place in the jungle where there is a lake or a river, and build their nests all around the lake, or by the bank of the river. The nest is just a heap of mud raised up from the ground, with a hollow at the top where the mother bird lays her eggs. Sometimes many thousands of flamingos are found together around one place, which is then called a *flamingo colony*. [175]

A Flamingo Colony
Photograph of a group in the American Museum of Natural History, New York

[177]

The flamingo is a very tall bird, taller than a man when standing up. The flamingo's legs are long and thin, and the neck is also long. The long neck and the long legs are very useful to him. He stands in the water on his legs, which look almost like a pair of stilts; then he bends down his long neck, dips his beak into the water, and catches a fish or any other small creature that he can find there. And although the fish or the small creature sees the flamingo's legs in the water, it does not run away. Why? Because it mistakes the legs for reeds growing in the water!

When thousands of flamingos in a colony are standing around the lake or by the river, where they live, it is a very grand sight from a distance. The flamingo's feathers are a bright red in color, with white or pink at the edges; so the thousands of flamingos look like an army of soldiers with red coats.

In former years, when soldiers sometimes wore red coats, travelers who happened to come toward a lake in Africa would suddenly see at a distance an army of soldiers, as they thought, standing by the lake. What they really saw were the flamingos fishing!

But no traveler could get very near the flamingos, for they have *sentinels*! I have told you that the wild buffaloes have sentinels to warn them when an enemy is coming. The flamingos have the same. Their sentinels stand here and there just outside the place where the others are fishing; and they keep a [178] lookout all the time. If any enemy comes, they cry out,

"Honk! Honk! Honk!" That means, "Enemy coming! Fly away!"

And of course all the flamingos rise up in the air and fly away to a safe place, till the enemy goes away.

To see a whole flock of flamingos flying in the sky far above one's head is a most wonderful sight. You have seen a cloud at sunset shining with lovely tints of red and pink and orange: well, the flock of flamingos flying in the sky looks something like that. And they all *keep level* at the same great height, in *rows and ranks*, just like an army, as there are thousands and thousands of flamingos in the flock.

Sometimes the rows and ranks widen out for a few minutes, and fill a large portion of the sky; then they close up again, and look like one long banner of red floating in the sky.

And all the time they have sentinels that fly outside the rows and ranks. They make the pattern in the sky still more beautiful.

My dear children, the more you think of these wonders of the jungle and of the world, the more you will understand how great and wise is God, Who made all these things. [179]

The Parrot

Another bright bird that lives in the jungle in a flock is the *parrot*. You know all about him, as you must have often seen him caged, or chained by the leg to a stand. But he is different in his happy home in the jungle. He lives in almost every sunny country, and flies about in flocks.

Wild parrots also make their nests in flocks. In India there is a deserted city called Amber. Once upon a time a great King lived there in a lovely marble palace; and the nobles and courtiers also had lovely marble palaces and mansions.

But one day the King said that they must all leave that city, and go and build another city. So everybody left the city of Amber, and to-day it still stands perfect—lovely marble palaces and mansions, with hundreds of bushes of wild roses growing all around them.

Nobody lives there, except thousands and thousands of wild parrots, that have made their nests upon the roofs of the palaces, in the porticoes and balconies, and upon the tops of the marble pillars and columns.

Just think of that lovely sight! The blue sky above, the red roses on the ground below, [180] and the white marble palaces between the blue sky and the red roses; and many thousands of green parrots flitting across the sky, and from palace to rose bush. Broad bands of red, white, and blue, with bright flashes of green between them!

Another lovely sight is a flock of wild parrots in the jungle, going home to roost at sunset. I once saw that sight. Their beautiful green wings and the patches of yellow on their heads shone amidst the gorgeous colors of the sunset. And as the parrots flew on and on, many thousands of them, their own colors mingled with the colors of the sunset in ever-changing bands. They flew toward the setting sun, and passed out of sight right into the sun, as it were.

After seeing a sight like that—seeing God's lovely creatures flying about like happy children at play—who wants to see a bird boxed up in a cage?

The Cockatoo

Another bright bird which you may have seen in a cage, or chained to a stand, is the *cockatoo*. He is a cousin of the parrot, but much larger, and far more gorgeous. He has a beautiful *crest* of red and orange feathers on his head. His [181] wings are a rosy pink in color; and he has a long pink and white tail.

In other ways he is very much like the parrot. He lives chiefly in the countries of southern Asia, and in the islands between Asia and Australia.

The Peacock

And now I come to the most beautiful bird of all, the *peacock*. When he spreads out his long tail, it looks just like a lady's fan, only far lovelier than any fan made by men. In color the tail is a kind of blue and green, with touches of gold and violet, and with "eyes" dotted all over it in shades of many other colors.

The peacock can also close up his tail like a fan. Then the long feathers of the tail all come together in many folds, and stand out a yard long behind him.

The peacock is found wild in India and in countries near there, but has now been brought into America and Europe. You may even have seen the peacock in the parks and gardens of some cities, where he lives quite peacefully, at least in the summer months. In the winter, of course, he must have a warm place indoors. [182]

The peacock is really the Papa bird, and the Mamma bird is called the *peahen*. She has not the gorgeous tail and the lovely feathers that he has; so she looks quite plain. You will find that *among animals the Papas are often much prettier than the Mammas.*

That seems very strange, does it not? Among us, of course, the Mammas are always prettier than the Papas!

But in another book I shall explain *why* the Papas among animals are often prettier than the Mammas.

The Golden Pheasant

There is another beautiful bird which has been brought to America, and now lives here; it is the *golden pheasant*. Once upon a time he lived only in China; but a few years ago people brought a number of golden pheasants to America, and put them in the forests of Oregon and Washington. So now there are many thousands of golden pheasants flying about and making their nests there.

There are other kinds of pheasants in England and in some parts of Europe, and these the people shoot and eat. But the golden pheasant is much too beautiful to eat. His feathers are as [183] lovely as the sunset—gold and yellow and orange, with blue and deep crimson; and all these colors are laid out on his feathers in such a beautiful pattern that to look at him you would think you were dreaming, and not looking at a real bird.

Man, who toils with his hands, cannot make such lovely colors as those of the birds of the air, and of the flowers in the fields, which do not toil.

The Snowy Egret

And now, my dear children, I shall finish this chapter by telling you about a beautiful bird that once lived quite wild in great numbers in the United States. This bird has lovely soft feathers, which are pure white; so it is called the *snowy egret*. The feathers are as soft as silk. They are also long, with a gentle droop at the end.

Because these feathers are so lovely, rich women want to wear them in their hats; and these rich women are willing to pay a great deal of money for the egret feathers. So, for the sake of the money, hunters go wherever these lovely birds are to be found, and catch and kill them, and get the feathers. In fact, they have [184] killed off so many of these lovely birds, to get feathers for rich women's hats, that to-day there are hardly any snowy egrets left in the United States.

Worse than that, the hunters killed the Papa and Mamma egrets just when their babies were born, because at that time the feathers of the snowy egrets were the softest and loveliest. And so, for each Papa and Mamma egret which the cruel hunters killed, they left a dozen *baby birds in their nests to starve and die*. Think of that!

Now, my dear children, I want the little girls among you to remember this, especially the little girls who are lucky enough to have rich Papas and Mammas. You can grow up to be beautiful, and look beautiful, without wearing these egret feathers. There are women who try to look beautiful, but who do not think of the pain they give to God's innocent creatures.

So, if ever you want to wear egret feathers, think of the dozen baby egrets who must starve and die if you are to have them. [185]

Snowy Egrets
Photograph of a group in the American Museum of Natural History, New York

[187]

CHAPTER XVI

The Caged Parrot

I shall finish this book by telling you a story—a true story, which, I hope, will make you think.

Many years ago a sea captain returned to his home in the north of Scotland, after sailing the sea for a long time. He brought with him a parrot. The parrot had lived in South America, where the people speak the Spanish language. So all the words the parrot knew were in Spanish.

The captain knew Spanish quite well, and often talked to the parrot in that language. But after a time the captain died, and there was nobody in that part of Scotland who could talk to the parrot.

The parrot grew silent, and never opened his mouth to say a word. But he was thinking of his friend who was dead, and whose words in Spanish had reminded him of his sunny home. The people around him did not know that, [188] and thought nothing of his silence. So the parrot in his cold and bleak cage pined and pined for his sunny home land, but never a word did he say.

Forty years passed, and a new set of people came to live there. They took no notice of the silent old parrot. They put food and drink in the cage, but knew nothing about him except that he had been in the cage for many years. For a parrot lives much longer than a man—sometimes one hundred years.

One day a sailor came to the house. He had lived in South America, and knew Spanish. He saw the parrot sitting in his cage, all alone and silent, with his head bent down, and his beak on his breast. Then the sailor spoke to the parrot in Spanish.

The parrot looked up, as if he had awakened from a long, long dream. Something reminded him of the days of his youth, when he was a happy bird flying about over the sunny fields of South America. Then he remembered the language of his youth, which he had not spoken for forty years.

Suddenly he flapped his wings in joy, and spoke again. He spoke all the Spanish words he knew, one after another. He spoke to that [189] sailor as to a friend come to him from his own home land. He flapped his wings against the bars, and finally dropped to the floor of the cage, dead. He had died in the thought of his bygone happy days.

My dear children, I am closing this book with this story, because I want you to learn a great lesson from it: *be kind to all animals*.

I know that you would never willfully hurt any animal. But that is not enough. You may think that you are very kind to some creature, because you feed it and pet it; but all the same you may be very cruel, though you do not mean to be so.

You may think it is great fun to have a pretty bird in a cage. But is it any fun *for the bird*? How would *you* like to be shut up in a cage all your life, instead of playing about in God's free air and living in your own home? The bird wants to fly about and live in his nest in his own home land. Think of that when you wish to put a bird in a cage.

Children who are kind to all animals grow up to be men and women who are kind to other people. And it is only by being kind to others that we ourselves *deserve* to be happy and *are* happy. [190]

Remember all that I have said, till I come back and talk to you again in the next book. Then I shall tell you many more Wonders of the Jungle.

Till then, as they say in the Orient, God and His peace be with you!